# Praise for *Dr. Tatiana's Sex Advice to All Creation*

"Witty, racy, informed, entertaining, and instructive . . . Judson is a gifted writer, and her book helps further understanding."
—*Science*

"Judson's witty, well-researched vignettes are a guilty pleasure, natural history that goes down as sweet as a trashy tell-all."
—*Outside* magazine

"Captivating . . . An evolutionary biologist with interesting and amusing things to tell us."
—*The Wall Street Journal*

"Erasmus Darwin titillated eighteenth-century London with his poem 'The Loves of Plants.' He never knew the half of it. Dr. Tatiana knows how the other half loves, and it's much kinkier than anybody imagined. Never has science seemed more like daytime television."
—Matt Ridley, author of *The Red Queen: Sex and the Evolution of Human Nature*

"Whimsical, irreverent and illuminating . . . A most amusing and educative book on animal sex."
—*San Francisco Chronicle*

"More positions than the Kama Sutra—but don't try this at home!"
—Steve Jones, author of *The Language of Genes: Solving the Mysteries of Our Genetic Past, Present and Future*

"Long live Dr. Tatiana. Without her, what would creatures troubled by their bizarre sex lives do? . . . *Dr. Tatiana's Sex Advice to All Creation* is charming, hilarious and exhaustively researched."
—*Miami Herald*

"Human? Lucky you! You rate a delightful romp through the weird, wild world of animal sex, with a guide who really knows the, um, ins and outs. . . . Count your blessings, primate. Then read, learn, enjoy."
—Melvin J. Konner, author of *The Tangled Wing: Biological Constraints on the Human Spirit*

# Dr. Tatiana's Sex Advice
# to All Creation

# Dr. Tatiana's
# SEX ADVICE
## TO ALL CREATION

## Olivia Judson

A HOLT PAPERBACK
Metropolitan Books   Henry Holt and Company   New York

Holt Paperbacks
Henry Holt and Company, LLC
*Publishers since 1866*
175 Fifth Avenue
New York, New York 10010
www.henryholt.com

A Holt Paperback® and 🄋 ® are registered trademarks of
Henry Holt and Company, LLC.

Library of Congress Cataloging-in-Publication Data
Judson, Olivia.
    Dr. Tatiana's sex advice to all creation / Olivia Judson.—1st ed.
    p.  cm.
  Includes bibliographical references and index.
    ISBN-13: 978-0-8050-6332-5
    ISBN-10: 0-8050-6332-3
    1. Sex.   2. Sex differences.   3. Sexual behavior in animals.   I. Title.
HQ25 .J83  2002
306.7—dc21                                       2002019591

Henry Holt books are available for special promotions and
premiums. For details contact: Director, Special Markets.

Originally published in hardcover in 2002 by Metropolitan Books

First Holt Paperbacks Edition 2003

Printed in the United States of America

D 30 29 28

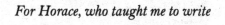

*For Horace, who taught me to write*

# CONTENTS

# AGONY AND ECSTASY:
# A NOTE FROM DR. TATIANA

In my business I get asked a lot of questions. Many of them concern matters beyond the wildest human imaginings. But the most common question is mundane enough: Why did I become a sex expert? Quite simply, I decided to dedicate myself to sex when I realized that nothing in life is more important, more interesting—or more troublesome.

If not for sex, much of what is flamboyant and beautiful in nature would not exist. Plants would not bloom. Birds would not sing. Deer would not sprout antlers. Hearts would not beat so fast. But ask an assortment of creatures, what is sex? and they will give you different answers. Humans and many other species will say copulation. Frogs and most fish will say the squirting of eggs and sperm in joint shudders of spawning. Scorpions, millipedes, and salamanders will tell you that sex is packets of sperm deposited on the ground for the female to sit on so they'll explode into her reproductive tract. A sea urchin will say sex is releasing eggs and sperm into the sea in the hope that they will, somehow, find each other in the waves. For flowering plants, sex is trusting the wind or an insect to carry pollen to a receptive female flower.

To succeed, each of these methods requires a suite of different features. A male flower who wishes to be a Lothario and have his pollen strewn to as many mates as possible must seduce not female flowers but bees. Other creatures must wear gaudy costumes, be they fancy feathers or frivolous fins; they must sing and dance for hours and hours; they must perform prodigious feats, building and rebuilding nests and bowers. In short, they must expend enormous energy shouting, "Choose *me*, choose *me*." And all for—what?

In truth, these various practices are just the means to an end. The ultimate sex act—the act that all these antics have evolved to accomplish—is the mixing of genes, the creation of an individual with a new genetic makeup. To a miserable organism sitting alone in a singles bar, genetic mixing might not seem worth the bother. Yet it is fundamental to the grand scheme of things. To see why, let's take a step back and think about how evolution works.

For most of us, caught up in the hurly-burly of our daily struggles, the purpose of life may seem elusive. But from an evolutionary point of view, the purposes of life are clear: survival and reproduction. If you fail at either, your genes go with you to the grave. If you succeed at both, you pass your genes on to your children. Inevitably—such is life—some organisms do better than others at surviving and reproducing. If everyone had identical genes, then differences in survival and reproduction would be due to luck, not genes. But usually individuals have different genes. And insofar as a particular gene confers an advantage in terms of survival or reproduction, that gene will spread.

This simple process, discovered by Charles Darwin and Alfred

Russel Wallace in the nineteenth century, is the principal mechanism of evolution. It is known as natural selection. Sometimes the process is fast and easy to see. Suppose a poison—an antibiotic, say, or an insecticide—appears in the environment. And suppose that surviving the poison depends on having a particular gene. Those who do not have the gene will die, their genes "deselected" from the population. In the extreme case, no one has the resistance gene, everyone dies, and the population becomes extinct. More often than not, however, some individuals are fortunate and have a gene to resist the poison. Since these individuals are the only ones to survive and reproduce, the genetic makeup of the population will shift to one where everybody is resistant.

Thus, genetic variation is crucial: no genetic variation, no evolution. But where does genetic variation come from? There are two main sources: mutation—and sex. Mutation, or random changes to information contained in genes, is the more primitive of the two. Mutations arise from errors made by the cell's genetic copying machinery. Since no scribe is perfect, some errors are inevitable—which is just as well. Whereas sex produces new combinations of genes that already exist, mutation creates altogether new genes—and thus generates the raw material of evolution. Without mutation, evolution would grind to a halt.

Yet mutation by itself is not enough. From time to time organisms evolve to give up sex, reproducing asexually instead. When this happens, any genetic differences between a parent and child are, by definition, due to mutation only. At first, asexual organisms often flourish. But their glory is fleeting. For reasons that remain mysterious, the loss of sex is almost always followed by swift extinction. Apparently, without sex you are doomed.

Which is not to say that sex makes life easy. No matter how good your survival skills are—you can be the champion at evading predators, or have the best nose for finding food, or be immune to every disease—it will all be for naught if you cannot find, impress, and seduce a mate. Worse, success at seduction is often at odds with survival. If you're a bird, flaunting an enormous tail may make you quite the cock among hens—but it may also make you lunch for a cat. Worse still, the competition for mates is often exceedingly stiff.

The upshot of all this is that the need to find and seduce a mate is among the most powerful forces in evolution. Perhaps nothing in life generates a more ecstatic diversity of tactics and stratagems, a more surprising array of forms and behaviors. In comparison, tricks to avoid predators seem predictable and limited. They typically include one or more of the following features: going about in groups, moving fast, blending in with your surroundings, looking scary, sporting a shell or sharp bits, or tasting revolting. But tricks to seduce a partner—ah, here the variety is endless. Which is why everyone asks so many questions.

And which is why I have dedicated my life to answering them. In the pages that follow, I have chosen samples of my correspondence over the years. I have deliberately selected questions that seem to me to address the concerns of all creatures, such as promiscuity, infidelity, and homosexuality. I have grouped questions on related topics into chapters, each of which has the briefest of introductions and concludes with a summary of my advice.

The chapters, in turn, fall into three related parts. In the first part, "Let Slip the Whores of War!" I reveal the reasons that males and females so often want different things from each other, and from life, and I explore some of the consequences. The second part, "The Evolution of Depravity," discusses situations where the collision is at its most intense—sometimes generating horrific outcomes, including rape and cannibalism. This part finishes with a brief chapter on the rarest, most aberrant evolutionary phenomenon: monogamy.

The last section goes further still, to ask "Are Men Necessary?" Here, I consider various matters to do with the evolution of gender and of sex. In an attempt to find out why sex is so essential for long-term evolutionary success, the final chapter is a discussion with the only organism so far known to have succeeded in living for millions and millions of years without it.

What of my methods? To respond to my readers, I have sifted through the scientific literature, reading hundreds of books and papers; I have interviewed specialists on subjects from dwarf males to giant sperm. When the answer is not known to science, as often it is not, I have given my best guess based on the knowledge available and my understanding of natural selection. Sometimes, as a result of my research, I have come to conclusions that are different from the prevailing wisdom; thus, I hope that in a small way this book makes a contribution to ongoing debates. In that spirit of open scientific inquiry I have provided, at the end of the book, all my sources.

I have found from experience that most organisms prefer to be called by their common names rather than by their technical Latin names—after all, few humans talk of themselves as *Homo sapiens*—so I use Latin names only when necessary for clarity or when the organism I am discussing is too obscure (or too snobbish) to have a common name. In observance of scientific custom, I have given measurements in metric units. However, in those instances where I felt it would be helpful to my readers in North America, I have provided conversions. Finally, I would like to thank my correspondents for agreeing to let me make public problems of a most intimate nature. This book would not have been possible without them.

# PART I

# LET SLIP
# THE WHORES OF WAR!

The battle of the sexes is no myth. Success at sexual
reproduction is at the heart of the evolutionary
process. But greater success for her often means
less success for him. The upshot? An eternal war—
and an astounding diversity of strategies.

# · 1 ·
# A SKETCH OF
# THE BATTLEFIELD

---

Boys are promiscuous and girls are chaste, right? Wrong. The battle of the sexes erupts because, in most species, girls are wanton.

---

*Dear Dr. Tatiana,*

*My name's Twiggy, and I'm a stick insect. It's with great embarrassment that I write to you while copulating, but my mate and I have been copulating for ten weeks already. I'm bored out of my skull, yet he shows no sign of flagging. He says it's because he's madly in love with me, but I think he's just plain mad. How can I get him to quit?*

*Sick of Sex in India*

Who'd have thought a stick insect would be among the world's most tireless lovers? Ten weeks! I can see why you've had enough. Twiggy, your suspicions are half right. Your paramour *is*

mad, though not with love but with jealousy. By continually copulating he can guarantee that no one else will have a chance to
get near you. It's a good thing he's only half your length, so he's
not too heavy to carry about.

Is your case unusual? Well, it is extreme, but by no means
unique. Males in many species are fiercely possessive of their
mates. Look at the Idaho ground squirrel, a rare critter that lives
only in (surprise) Idaho. The male won't let his partner out of his
sight and follows her everywhere. If she goes into a burrow he
sits at the entrance so she can't come out—and no one else can go
in. Worse, he picks fights with any other male who happens to
come near. Or consider the blue milkweed beetle. After sex—
which by any insect standards is brief, taking only ten minutes or
so—the male insists on riding around on the female's back, not
so he can whisper sweet nothings but so he can stop her from
gallivanting.

To be frank, though, males have good reason to be possessive.
Given half a chance, girls in most species will leap into bed with
another fellow. "But wait," I hear you cry, "isn't it a general law of
nature that males are promiscuous and females are chaste?" That
is indeed what used to be thought. But now we know this notion
is nonsense.

The man who first lent scientific respectability to this notion
was named A. J. Bateman. In 1948, he published a paper in the
scientific journal *Heredity* in which he claimed to have proved that
males have evolved to make love and females to make babies. His
assertions were based on experiments he had conducted with the
fruit fly *Drosophila melanogaster*. This is one of those small flies
drawn to bowls of ripening fruit or to glasses of wine, and should
on no account be confused with the larger housefly, which likes
feces and other muck. *Drosophila* boasts perhaps as many as
two thousand different species, with more than four hundred in

Hawaii alone. Little is known about most of them. But *Drosophila melanogaster* is beloved of geneticists and, along with the worm, the mouse, and the human, is one of the most studied animals on earth.

After keeping equal numbers of adult male and female flies together in small bottles for three or four days, Bateman noticed that the males were keen to mate as often as possible, ardently vibrating their wings for any girl who'd pay attention. If a female was receptive, the male would press his suit, licking her genitalia before gently spreading her wings and mounting her. Most of the time, however, the males were disappointed. Bateman observed that female flies tended to reject the advances of more than one or two suitors. In accordance with this observation, he also found that while males had more children the more partners they mated with, females did not.

To explain his results, Bateman appealed to what he saw as the essential difference between the sexes—that males produce lots of tiny, cheap sperm whereas females produce a few large, expensive eggs. He also noted that the females of many species can store sperm for days, months, or in some cases years, which means that sperm from one mating can, in principle, last a lifetime. Therefore, Bateman argued, one male could easily fertilize all the eggs of many females. Accordingly, he went on, females are limited in their reproduction by how fast they can make eggs, whereas males are limited only by the number of females that they can find and seduce. So, he proclaimed with a flourish, males (including humans) are natural philanderers while females (again including humans) are naturally chaste—and in all but the most unusual circumstances should be indifferent or hostile to mating more than absolutely necessary. From this perspective, Twiggy, your mate's passion is weird and inexplicable: he should be off seducing other stick insects, not sticking stubbornly to you.

Nonetheless, this "men are cads, women are saints" hypothesis—more politely known as Bateman's principle—has been all the rage. Patriarchs have extolled it. Feminists have invoked it. Scientists have expounded it and buttressed it, adding extra reasons—the risk of venereal disease or being caught by a predator while copulating—why females would naturally want to keep sexual activity to a minimum. And certainly there are some species—such as the alfalfa leaf-cutter bee—where females mate just once. There are other species where males race from one girl to the next or are so eager they'll fornicate with anything: goldfish, for example, are occasionally drowned by amorous frogs. But as a general rule? Ha!

Bateman's principle has a fundamental flaw: it's wrong. In most species, girls are more strumpet than saint. Rather than just mating once, they'll mate with several fellows, and often with far, far more than necessary just to fertilize their eggs.

How did Bateman come to be so misguided? There are two reasons. The first is a quirk of fate. As I said, *Drosophila melanogaster* was—and remains—the most fashionable fruit fly to study. Females of this species really are quite restrained, preferring not to mate more than once a week or so. Other species of *Drosophila* would have given different results: *Drosophila hydei* females enjoy sex several times each morning, for example. But even in *melanogaster*, females are not as virtuous as Bateman thought. The trouble is—and this is the second reason he blundered—his experiments were too short. We now know that if he had continued them for one more week, he would have discovered that *Drosophila melanogaster* females recover their appetites—and indeed that those who mate only once have fewer children than their more lascivious sisters.

The reason it took more than thirty years for anyone to detect the problems with Bateman's theory is partly because his logic sounded reasonable. Moreover, it seemed to be supported by observation. Thousands and thousands of hours spent watching mammals and birds going about their lives had given no suggestion that females were often unfaithful to their mates. That's not the whole story, though. Even once scientists started to notice that females in some species—especially insects—do mate with a lot of males, they didn't immediately grasp the full implications. If females mated more than expected, it was assumed they had "malfunctioned" or that males had led them astray, not that females might have something to gain.

In the 1980s, the development of more sophisticated genetic techniques meant that biologists could find out who was really having whose children. And they discovered something astonishing, something that no one had predicted. Namely that, from stick insects to chimpanzees, females are hardly ever faithful.

This discovery was swiftly followed by another that was even more surprising: in species after species, rampant promiscuity is no malfunction. Rather, females benefit from it. My files bulge with examples. To pick a few at random, female rabbits and Gunnison's prairie dogs both show higher rates of conception if they mate with several partners while they are in heat. The female sand lizard hatches out more eggs the more lovers she's had. The female slippery dick—a pale fish that lives on coral reefs—will have more of her eggs fertilized if she spawns with a gang than if she spawns with just one fellow.

These discoveries have forced a reevaluation of male and female behaviors that is still under way. But one conclusion is inescapable. As we shall see, when females mate with more than one male, War capers nimbly through the boudoirs, imps of discord frolicking in his wake.

∽

*Dear Dr. Tatiana,*

*My boyfriend is the handsomest golden potto I ever saw. He's got
beautiful golden fur on his back, creamy white fur on his belly, he
smells delicious, and he has ever such dainty hands and feet.
There's just one thing. Please, Dr. Tatiana, why is his penis covered with enormous spines?*

*Spooked in Gabon*

All the better to tickle you with, my dear. At least, I'll bet that's
a big part of the reason. Golden pottos are little-known relations
of bushbabies—small, night-climbing primates that are distant
cousins of monkeys and apes. If you look at your cousins, you'll
see your beloved is not alone. Bushbabies and many other primates have monstrous penises—many of them look like medieval
torture instruments. They have spikes and knobs and bristles and
are often twisted into weird and sinister shapes. By comparison,
the human penis is dull, notable only for its girth.

Penises are for more than just sperm delivery, you see. If
females mate with a number of males, each subsequent suitor will
sire a larger proportion of her children if his sperm are the ones
that do the trick. A male who can stimulate his mate to take up
more of his sperm, or who can somehow get rid of the sperm of
his rivals, will spread more of his genes than less artful fellows.
Thus, the first consequence of female promiscuity is that males
are under great pressure to outdo one another in all aspects of
love. For this task, the penis is an important tool.

Consider damselflies. These insects, close relations of dragon-
flies, look sweet and innocent as they flit along the riverbanks
on a sultry summer day. But they have evolved some of the fanci-
est penises around. A typical damselfly penis has a balloon—an

inflatable bulb—and two horns at the tip, plus long bristles down the sides. In the black-winged damselfly, *Calopteryx maculata*, the male uses this device to scour sperm from inside a female before depositing his own. But in the related *Calopteryx haemorrhoidalis asturica*, he uses his penis as an instrument of persuasion: by stimulating her in the proper manner, he can induce her to eject sperm from previous lovers. Meanwhile, the moth *Olceclostera seraphica* has genitals that resemble a musical instrument: the male rubs one part of his privates against another, producing vibrations with which to thrill his mate. In contrast, among termites the female typically mates with only one male—and male termites have plain, unadorned genitalia that do not differ much from one species to the next.

A penis is not the only way to outdo other males, of course. Take the ghost spider crab *Inachus phalangium*, a creature that lives under the protective tentacles of sea anemones. The male makes a special jelly to seal the sperm of previous males into a corner of the female's reproductive tract so it won't be able to mingle with his own. Or consider the dunnock, a bird that looks like a sparrow that has fluffed its feathers in ashes. Most male birds—swans, ducks, and ostriches being exceptions—do not have penises. Instead, males and females copulate by quickly pressing their genital openings together. Hardly satisfactory. But even without a penis, male dunnocks have found a way to get rid of rival sperm. Before sex, the male pecks the nether regions of his mate; sometimes this encourages her to dump any sperm she's collected. Even more exotic: the red-billed buffalo weaver, an African bird that lives in small communities. The female is wildly promiscuous. Apparently in response, the male has evolved a pseudophallus—a rod of tissue that cannot transfer sperm. During sex, he rubs this rod against the female's genitalia for about half an hour—at which point he ejaculates from his genital opening and appears to have an intense orgasm. The male who provides the most vigorous stimulation is presumably the most successful at persuading a female to use his sperm.

All of this provides a clue as to why your friend's penis looks so alarming. Among primates as among insects, it is a rule of thumb that in species where females consort with one male at a time, penises are small and uninteresting. I mean, take the gorilla—a huge guy with a little teeny weenie. A male gorilla can weigh 210 kilos (460 pounds), but his penis is a measly five centimeters (two inches) long and entirely devoid of knobs and spikes. The Argentine lake duck puts him to shame. The duck is small, but his penis, which rivals that of the ostrich, is twenty centimeters (eight inches) long—and it has spines. But then, a male gorilla generally presides over a small group and does not often have to worry about other fellows' sperm. If I were a girl gorilla, though, I'd feel I was missing out: as far as anyone can tell, the females of more promiscuous primate species are more capable of orgasm. So I'd guess your mate's penis is gloriously bespined because female golden pottos sometimes sleep around. But whether the spines have evolved because you like them or whether they are more for scrubbing—well, why don't you find out?

∽

*Dear Dr. Tatiana,*

*I'm a queen bee, and I'm worried. All my lovers leave their genitals inside me and then drop dead. Is this normal?*

*Perplexed in Cloverhill*

For your lovers, this is the way the world ends—with a bang, not a whimper. When a male honeybee reaches his climax, he explodes, his genitals ripped from his body with a loud snap. I can see why you find it unnerving. Why does it happen? Alas, Your Majesty, your lovers explode on purpose. By leaving their genitals inside you, they block you up. In doing so, each male hopes you will not be able to mate with another. In other words, his mutilated member is intended as the honeybee version of a chastity belt.

You may think this is no way to treat a queen. But even queens are not exempt from the battle of the sexes. Indeed, I'm afraid your situation appears to exemplify the full, complex, and dynamic conflicts of interest that can arise as a consequence of female promiscuity.

To see how the conflict unfolds, let's first look at matters from the male's point of view. His plight is desperate. A young queen such as yourself spends a mere few days mating before going off to start a nest. After that, you'll never bother with sex again; you'll be too busy having your half million children. Worse, his chances of mating with you are small to start with. Honeybees have sex on the fly: you take to the skies and mate with any male who can catch you. The competition can be fierce: as many as twenty-five thousand males may assemble to contend for a single queen. But you probably won't mate more than twenty times, so most male honeybees die virgins. Any male who succeeds in catching you has nothing to lose by exploding: he'd be unlikely to mate again anyway. What's more, he may have something to gain. If, by blocking you up, he can prevent just one other male from copulating with you, he will fertilize a larger proportion of your eggs—and more of his genes will be passed to the next generation.

But the problem is, while his interests are best served if you mate only with him, you do better if you mate with several males. Indeed, a queen who mated only once could be at risk of losing half her brood. Why? Because of the complicated way that gender is determined in honeybees.

Usually, male honeybees hatch from unfertilized eggs, females from fertilized eggs. But bees have a gene, known as the sex-determining gene, that can mess up this arrangement. If a queen mates with a male who has the same version of this gene as she does, then half her fertilized eggs will hatch out sons—and sterile sons at that. Thus, instead of producing a mixture of dutiful daughters working hard to rear their sisters and a few fertile males waiting for their once-in-a-lifetime opportunity to explode with joy, half her children would be good-for-nothing infertile males—which the dutiful daughters would eat alive. Such a reduction in the workforce increases the risk that the nest will fail. If, therefore, the queen mates with several males instead of just one, any male whose sex-determining gene matches hers will fertilize a smaller proportion of her eggs. This way, only that small proportion of her offspring will be sterile males. So the more a queen mates, the more likely she is to avert disaster.

That's not all. Males, obviously, will also gain if they can thwart previous lovers by removing the plug and mating with the female in their turn. You might imagine, then, that male honeybees would have evolved some way of removing the chastity belt. You'd be right. If you look closely, you'll see that each male honeybee sports, on the tip of his phallus, a hairy structure that can dislodge the severed genitalia of his predecessor.

This suggests the following evolutionary scenario. Once upon a time, queens mated with only one male. Then a mutant queen appeared who mated with more than one. She was more successful at reproducing than her virtuous sisters, and the gene for multiple mating spread throughout the honeybee population. Then a male appeared who, by exploding, prevented the queen from mating again. Genes for exploding males spread throughout the population. In a counter-countermove, the queen evolved to block the male's advantage, either removing the plug herself or perhaps having it removed by the workers (this step would have happened swiftly, since any female who did not remove the plug would not have been able to lay eggs). Then males evolved their own counter-counter-countermove. And so on.

As you've probably guessed, such a situation is far from unusual. It is generally the case that whenever females mate repeatedly, males are sure to lose. Any male who can prevent a female from mating with his rivals will sire more of her children than a less controlling fellow and will thus spread more of his genes. So you shouldn't be surprised to learn that chastity belts are a popular evolutionary invention, in vogue among bats, rats, worms, snakes, spiders, butterflies, fruit flies, guinea pigs, squirrels, chimpanzees—I could go on. I must admit, however, that most of these fellows opt for the more traditional plugs, cements, and glues rather than for amputating their genitalia. In many species of rodent, for example, males have enormous glands to secrete tough, rubbery corks that they place deep in their partners' reproductive tracts as they finish copulating. The house mouse makes a plug so tough that a scalpel virtually bounces off it; once the plug has formed inside a female, attempts to remove it can tear the ligaments of her womb.

But alas, poor males. Whenever repeated mating is beneficial to the female, she will gain by resisting male efforts at control. Thus, as males evolve to control, females evolve to resist. Which is why not all chastity belts are as effective as they might be. A female fox squirrel, for example, reaches around and pulls out the plug right after sex (and sometimes then eats it—how delicious). Moreover, males will also be under pressure to evolve to remove the plug. Again, this skill has appeared repeatedly. In the rat, the male's penis is almost prehensile: it can do some glorious gymnastic flips to dislodge plugs left by previous lovers, making like a toilet plunger and pulling them out with suction.

So you see, the battle of the sexes is fought on two fronts. Conflicts of interest between males and females mean that every new weapon or behavior evolved by one sex will favor traits in the other sex that can counter that development. At the same time, males evolve to manipulate and thwart a female's previous and subsequent lovers. If you watch through the generations, you'll witness a mighty evolutionary battle.

ᆼᄉ

Men, you're in a cruel bind. Female promiscuity puts your genes at risk: it's no good seducing all the women in sight if none of them uses your sperm. A woman's potential for promiscuity curbs your own and exerts a powerful force on your evolution. Rather than maximizing the number of girls seduced—that is, acting like a cad—you should try to maximize the number of eggs fertilized. For some men, some of the time, this will amount to the same thing: more skirt chasing. On many occasions, though, cads who spread their bounty will have fewer offspring than more loyal fellows, and so genes for pure caddishness will decline in frequency. Far better would be to attach yourself like the stick insect, explode like the honeybee, or evolve still other fates stranger than your strangest dreams.

# ·2·
# THE EXPENSE IS DAMNABLE

It's not easy being a male. Especially if you have to make sperm twenty times longer than you are. Or produce billions and billions of sperm in every ejaculate. Or copulate a hundred times a day to satisfy your partner. Or perform some other feat of prodigious sexual prowess.

*Dear Dr. Tatiana,*

*I'm a splendid fairy wren, and I'm concerned about my husband. He keeps going to the doctor because he's convinced his sperm count is too low and we won't be able to have children. But he ejaculates eight billion sperm at a time, so I don't see how he can have a shortage. Has he really got a problem, or is he being neurotic?*

*Bewildered Down Under*

The doctor, eh? I'd say your mate is not a hypochondriac but a liar. His "appointments" are a thin disguise for philandering.

Splendid fairy wrens are notorious for their extramarital adventures. Let me give you a tip. You can tell when a fairy wren's going a-wooing: he'll be carrying a pink petal in his beak to offer to his paramour. Why pink? It looks pretty when he fans his iridescent blue cheek feathers.

But the real question is why a bird smaller than my fist should need an ejaculate containing more than eight billion sperm. In human beings, the average dollop contains only 180 million or thereabouts. Even that's odd when you think about it, though. All those sperm for one little egg. Why?

Sperm numbers are a crude index of the difficulty of reaching an egg. If you're a tree, for example, the amount of pollen you make depends in large part on how it gets delivered. Take fig trees. Some species are pollinated by virtuous wasps, wasps that actively collect and distribute pollen; these figs can afford to be frugal with the stuff. Other fig species are forced to rely on lazy wasps that just brush against the flowers; these figs are, perforce, profligates. So in species like yours or mine, where a male delivers his sperm personally, you would imagine that sperm counts would drop.

Not necessarily. In some species—of fish, for example—males and females meet up, but instead of copulating, they spawn eggs and sperm into the water. And yes, in such cases sperm do not dramatically outnumber eggs. But when you turn to birds, mammals, and other groups that copulate, the picture changes. Instead, you find that males make the most sperm in species where—drumroll, please—females are sluts.

Two factors are thought to produce massive sperm counts in species where the girls are promiscuous. The first is what biologists call "sperm competition"—that's right, sperm from different males vie with each other to fertilize eggs. If sperm competition proceeds according to a raffle—the more tickets you buy the more likely you are to win—then the males that produce the most sperm will have the best chance of fertilizing a female's eggs. And if the number of sperm they make has a genetic basis, then over time the repeated success of males with the highest sperm counts will lead to an increase in the number of sperm each male produces. Reflecting this, males regularly exposed to sperm competition should also have larger testes—sperm factories—for their body size. Indeed, an experiment with yellow dung flies—hairy flies that mate and lay eggs on fresh cowpats—has shown that testes size can evolve in response to sperm competition in as few as ten generations.

Carrying the argument to its logical conclusion, then, males who are never at risk of sperm competition should produce only enough sperm to fertilize each egg. Unfortunately, few fellows are in this luxurious position. But males in one group are—seahorses and their close cousins, pipefish. (Pipefish look like seahorses that have been straightened and streamlined.) These males are famous for their pregnancies. Females typically deposit eggs in the male's brood pouch, and he fertilizes them there—so there's no chance of his sperm encountering those from another male. Most species of seahorse have not had their sperm counted. But one that has is the seaweed pipefish, a species common in seagrass beds around Japan—and sure enough, his sperm count is vanishingly low.

The second factor thought to produce high sperm counts is that sperm die in huge numbers on their travels through the female reproductive tract: of the millions that set out, just a few succeed in reaching their goal. Fantastically high sperm death has been remarked upon for more than three hundred years, yet there is still no good theoretical explanation for why so many sperm die in the first place.

You see, the surprising fact is that female reproductive tracts are often hostile to sperm. No one knows why this is so. But rather than cosseting sperm and helping them on their way as you might expect, reproductive tracts are full of hazards and treachery. Sperm may be digested, ejected, or rounded up and removed. Even in species where the females store sperm for years, they only keep a few of the sperm that they get. A honey-bee queen who mates with seventeen lovers, for example, will receive an average of 102 million sperm (6 million from each fellow); yet she'll keep a mere 5.3 million to fertilize her eggs. In species that don't store sperm—it's carnage.

In humans, for example, sperm start their odyssey in the acidic environment of the vagina. But acid is lethal to sperm (which is why strategically placed lemon slices make a good, if rudimentary, contraceptive), and fewer than 10 percent will continue their onward voyage. Survivors must then traverse the cervix, a barrier coated with mucus that even at the best of times permits a mere 10 percent of sperm to cross. Mucus is only one of the cervical hazards, though. At the first hint of sperm, white blood cells—the foot soldiers of the immune system—swamp both the cervix and the lining of the womb and destroy any intruders they encounter. In rabbits, within an hour of copulation an enormous army of white blood cells assembles at the cervix; in women, the army amasses within fifteen minutes of copulation, and within four hours it numbers more than one billion. By the time sperm reach the fallopian tubes—the place where any fertile eggs might be hanging about—their numbers will have dwindled from millions and millions to a few hundred. Which is why men with a sperm count of fifty million—which might seem generous—are likely to be infertile.

Measuring hostility is much harder than counting sperm, so we don't know how hostility varies among species or even among individuals of the same species. I would guess, however, that females respond to escalating sperm numbers by increasing the hostility of their reproductive tract. This in turn would favor males who produce more sperm. In the rabbit, for example, the number of sperm arriving at any given stage of a female's genital tract depends on how many sperm started out. But what do females gain from being hostile? After all, it seems counterproductive: if they are too hostile, their eggs won't be fertilized and they'll have no children. One idea is that hostility ensures that only the most superior sperm succeed in fertilizing eggs. Another is that hostility may initially evolve as a defense against infection. This would then be met by male attempts to bypass the defenses. Indeed, the semen from humans and many other mammals contains substances known to suppress the female's immune response. To counteract male attempts at suppression, females may respond by increasing the magnitude of their immune response—setting up an escalating evolutionary cycle of response and counterresponse.

Which brings me back to why your husband needs so much sperm. In splendid-fairy-wren societies, couples live together and rear children together—but free love prevails. Splendid fairy wrens are splendidly promiscuous, and most females have at least one lover as well as a husband. As a result, sperm competition is intense. Often, none of the chicks in a nest will have been sired by the male who is rearing them. So while your husband is off philandering, perhaps you've engaged in a little mischief of your own?

∽

*Dear Dr. Tatiana,*

*I've heard it's going to take me three weeks to make just one sperm. Apparently this is because it's going to need a tail twenty*

*times longer than my whole body. This seems awfully unfair:*
*I'm just a little fruit fly,* Drosophila bifurca. *Can't I get a*
*prosthesis?*

*Waiting for Sperm in Ohio*

There's no market in artificial sperm tails: you're going to have
to make them yourself. You're right—it's not fair. Why should a
fruit fly three millimeters long—smaller than this dash—have to
make sperm that measure fifty-eight millimeters? A human is far
bigger than you, but gets away with sperm one thousand times
smaller. Indeed, if a man were to make a sperm on your scale, it
would be as long as a blue whale. Now, that I'd like to see.

Unlike sperm number, the evolution of sperm size and shape is
something we know little about. All we can say for sure is that
sperm tend to be smaller and simpler in species where the eggs
are fertilized outside the female's body.

Consider sperm shape first. A conventional sperm looks like a
tadpole, with a big head and undulating tail. But in many species,
sperm deviate from this model. A popular invention is the tandem sperm—sperm that always swim in teams of two. They've
evolved in American opossums, water beetles, millipedes, firebrats, and some seafaring snails. Hooks are also fashionable.
Koalas, rodents, and crickets all have hooked sperm. The protura,
tiny critters related to insects, were the first to play Ultimate
Frisbee: their sperm are flat discs. Crayfishes have sperm that
resemble Catherine wheels. Some land snails make corkscrews.
Some termites make bearded sperm—sperm with one hundred
tails. Roundworms make amoeboid sperm: these don't swim, they
crawl. And all this before you begin to look at spermatophores,
the sperm packets that many creatures deliver. After a lengthy
lovemaking session, the giant octopus, for example, hands over a

spermatophore that is a huge bomb. Over a meter (three feet) long, it contains more than ten billion sperm and explodes inside the female reproductive tract.

Since these various shapes have evolved independently in different groups, they must be repeatedly beneficial in some way. Hooks, for example, might help sperm grapple their way along the reproductive tract—but to my knowledge, this has never been demonstrated. The possible advantages of other shapes? Your guess is as good as mine. But as far as we can tell, sperm shape has nothing to do with female promiscuity.

Sperm size, however, probably does. In roundworms, big sperm are more successful at fertilizing eggs, apparently because they crawl faster than small sperm and are less likely to be pushed aside by rivals. Similarly, in the bulb mite, an agricultural pest, males who boast big sperm fertilize more eggs than males whose sperm are puny. Indeed, as a general rule, where females are promiscuous, males don't just make more sperm, they make bigger sperm. Yet both attributes can't increase indefinitely: at some point, making sperm bigger must mean producing them in smaller quantities. In most species, pressure to keep the numbers up probably precludes a substantial increase in sperm size.

In a few species, though, numbers are apparently reduced in favor of making enormous sperm. The Giant Sperm Hall of Fame contains a diverse scattering of animals. Although you, *Drosophila bifurca*, are the reigning champion, in the last hundred years the title of Most Massive Sperm has been held by featherwing beetles, back-swimming beetles, ostracods (small shrimp that look like kidney beans on legs), ticks, the Australian land snail *Hedleyella falconeri*, the painter's frog, and many other species of fruit fly. The ostracod sperm, by the way, are rumored to fight, smashing one another to smithereens—although as best as I know this has never been investigated in a laboratory.

Unfortunately, work on giant sperm hasn't advanced much beyond openmouthed gawking, so I can't say a great deal about why some animals make gargantuan gametes. But what we do know suggests that big sperm have nothing to do with big eggs, although this idea has been considered. People don't study eggs as much as they study sperm (in species where fertilization is internal, sperm are easier to observe), but other fruit flies make both bigger eggs and smaller sperm than you do. Another idea is that giant sperm are a present from the male to the female to provide nutrition for her eggs. But in many species with giant sperm, only a tiny piece is taken into the egg, so I'm not convinced by this explanation either. Could giant sperm block the female reproductive tract and therefore serve as a chastity belt? This may be true for featherwing beetles: giant sperm from one male appear to fill the female's reproductive tract, effectively stopping the addition of sperm from anyone else. But this is not the case for ostracods: the females have a bizarre system of sperm storage whereby sperm are deposited and stored in an area that has no direct opening into the chamber where eggs are made. To fertilize an egg, an ostracod sperm must actually leave the waiting room and travel outside the female's body to reach the egg chamber. And in your closest competitor, *Drosophila hydei* (sperm 23 millimeters), not only does a female mate repeatedly but the various sperm she receives will mix. As a result, if she mates with several fellows on one day, each sires an equal proportion of her brood.

Still, there must be a good reason for colossal sperm. After all, where the cost of making them has been measured, it turns out to be considerable. Whereas your distant cousin *Drosophila melanogaster* (sperm 1.91 millimeters), can begin copulating a few hours after he tumbles out of his pupa, you must wait at least seventeen days—the time it takes for you to build your enormous testes. But things could be worse. If you stay out of trouble, you can expect to live for six months—a long time for a fruit fly—so having to wait seventeen days to lose your virginity isn't such a trial. In another of your cousins, *Drosophila pachea* (sperm 16.53 millimeters), a male spends the first half of his adult life unable to reproduce. And here's something you can feel good about: while most males need armies of millions, apparently you can thrive with a few good sperm.

∽

*Dear Dr. Tatiana,*

*I'm a furious fruit fly,* Drosophila melanogaster. *When I was a maggot, I was told that sperm were a dime a dozen, easy to make and easy to spend. So, on reaching maturity, I spent. With reckless abandon. But I was told a lie: I'm only partway through my life as a grown-up fly yet I've completely run out of sperm and no girls will come near me. Who can I sue?*

*Dried Up in London*

Our friend Bateman has a lot to answer for. "Sperm are cheap" was one of his notions. But it's the biggest myth in town. And although I'm sorry you've been misled, I can't help feeling a certain smug satisfaction that *Drosophila melanogaster,* Bateman's model organism, should turn out to have problems of this kind.

To recap: Bateman argued that because one sperm is cheaper to make than one egg, the factors limiting reproduction in males and females are different. Females, he said, are limited by eggs produced, males by mates seduced. According to this view, sperm are—to all intents and purposes—unlimited, and every single egg stands a good chance of being fertilized.

All too often this is not so. In the sea, animals from sponges to sea urchins do not rendezvous with their partners but release sperm into the water. Some species release eggs into the water, too. This means that sperm can have a tough time meeting up with eggs. In many such species, a high proportion of an individual's eggs may go unfertilized. Small wonder that some sponges spew sperm as if they were finalists in a Vesuvius look-alike contest, sending up large thick clouds of the stuff for anywhere from ten minutes to half an hour at a time.

On land, plants can have a similarly rough time. Pollinators—animal go-betweens, like bees—can be unreliable, preferring to eat pollen instead of delivering it to female flowers, so plants are often limited by the amount of pollen they receive. The cheeky forest flower jack-in-the-pulpit produces ten times more seeds if it is pollinated by a scientist rather than by a fly. Nor are such difficulties restricted to organisms who have to gamble on ocean currents or the wind or who are dependent on the vagaries of messengers. The lemon tetra, a small fish that lives in the Amazon, cannot fertilize all the eggs one female produces on a given day. His success at fertilizing falls the more he spawns, and he quickly gets to a point where his energy is better spent making more sperm than seducing more mates. No surprise, then, that female lemon tetras seem to prefer males they have not seen spawning with anyone else. In the bluehead wrasse, an Atlantic coral-reef fish, the biggest males husband their sperm, handing it out stingily—and often at levels below those best for females.

The trouble is, of course, most males are not producing just one sperm for each egg. They are producing hundreds, thousands, or millions of sperm per egg. That gets expensive. Male garter snakes have to rest for twenty-four hours after sex (which, to be fair, has usually been rather vigorous). A male zebra finch, a small bird with black and white stripes, who copulates three times in three hours, will run out of sperm—and it will take him five days to refill. Male blue crabs must wait fifteen days to replenish supplies. Even rams, who supposedly hold sperm reserves for ninety-five ejaculations (a typical man holds enough for one and a half) soon find their sperm counts going into freefall. After six days of sex, the sperm in a ram's ejaculate can fall from more than ten billion to less than fifty million—a threshold below which he'll have a hard time impregnating anybody. And some snakes positively fade away. The adder, a poisonous European species, loses a lot of weight at the start of the breeding season even though he's doing nothing but lying in the sun making sperm. That's one way to burn calories.

But the clinching proof that sperm are often limiting comes from hermaphrodites—organisms such as garden slugs and snails that are both male and female at the same time. Under the sperm-unlimited theory, hermaphrodites should run out of eggs before they run out of sperm and, given a choice, should prefer the male role over the female one. But in many species, this is not what happens.

Consider *Caenorhabditis elegans*, a tiny transparent round-worm beloved of geneticists. *C. elegans* is different from most hermaphrodites because it comes in two sexes: hermaphrodites and males. Among conventional hermaphrodites, there are two ways to have sex. Copulation can either be bilateral—both partners inseminate each other at the same time. Or it can be unilateral—in any given bout of sex, one partner plays the male role and the other partner plays the female. In *C. elegans*, however, the hermaphrodites cannot mate with each other, but they do make both eggs and sperm and can fertilize themselves. (Males, obviously, make only sperm.) A hermaphroditic *C. elegans* who never encounters a male will run out of sperm after laying about three hundred eggs. We know that the sperm are finished up first because for a while afterward the animal carries on laying unfertilized eggs—sometimes as many as a hundred.

But perhaps *C. elegans* is a special case. The hermaphrodite form does not make eggs and sperm at the same time; rather, it does sperm first. Therefore, the more sperm it makes, the longer it must wait before making eggs and the older it is when it starts reproducing. Too much delay is bad: among these worms, you're likely to have more offspring if you get off to an early start.

*C. elegans* and its relations are not, however, the only hermaphrodites that run out of sperm. Sperm limitation has also been found among sea slugs, sea hares, freshwater snails, and freshwater flatworms. (Although these organisms appear similar, they are only distantly related. Their looks and lifestyles have been arrived at independently.) In the freshwater flatworm *Dugesia gonocephala* (a bilateral copulator), sperm packets take two days to prepare, so individuals are thrifty with sperm and don't give away more than they receive: they stop giving sperm when their partner does. In the sea slug *Navanax inermis* (a unilateral copulator), individuals appear to prefer the female role over that of the male—the opposite of what you'd expect if sperm were unlimited.

And lest anyone still doubt that the male role can be expensive, look at banana slugs—enormous yellow slugs of the Pacific Northwest. In these hermaphrodites copulation is unilateral—and in several banana slug species, individuals may get only one shot at being male, whatever their sperm count. Banana slug penises are gigantic and complex. During sex, the penis often gets stuck. At the end of sex, therefore, the slug or its partner gnaws off the offending phallus. It never grows back: from that point on, the slug plays only the female role.

That said, let's take a closer look at your situation. *Drosophila melanogaster* males apparently suffer from two types of mating-induced sterility. The first is temporary: after mating once, a male ought to rest for a day before mating again in order to replenish supplies. The second type appears to be permanent. Unfortunately, because of the design of the experiments done so far, we don't know how quickly permanent sterility arises. All we know is that males presented with two females every two days are completely and irreversibly sterile by day 34—just under halfway through their adult lives. Perhaps in the wild, male flies never mate often enough—or live long enough—for this to be a problem. Perhaps. But then again, perhaps not. It is surely no accident that in your species, as in many others, females prefer to mate with fresh-faced virgins.

༄

Dear Dr. Tatiana,

My lioness is a nymphomaniac. Every time she comes into heat, she wants to make love at least once every half an hour for four or five days and nights. I'm worn out—but I don't want her to know. You couldn't suggest any performance-enhancing drugs, could you?

Sex Machines Aren't Us in the Serengeti

There is a drug, but it hasn't been approved for lions yet, I'm afraid. And anyway, you should be ashamed of yourself. A great big lion like you should be able to keep it up without fussing. I've heard of lions copulating 157 times in fifty-five hours with two different females. Honestly.

But let's look at the reason for your lady friend's prodigious lust. The problem is, she's got a genuine, clinical sex mania. Such manias are of two main types. In type I, the female needs lots of

stimulation to get pregnant. In type II, a male copulates like mad
not to stimulate the female but to ensure all the offspring in a
brood are his own. Your lover is a classic type I. The affliction is
not limited to lions: female rats, golden hamsters, and cactus
mice, for example, all require vigorous stimulation before they
can conceive. But lionesses are especially demanding: by some
estimates, less than 1 percent of all copulations produce lion cubs.
No wonder lions spend so much time snoozing.

What does all the stimulation do? Some species—such as
rabbits, ferrets, and domestic cats—do not release eggs without proper stimulation. Others—rats, for instance—release eggs
spontaneously, but a female that hasn't been stimulated enough
won't stay pregnant even if her eggs have been fertilized. And
lions? The usual assumption is that lions are like domestic cats
and won't ovulate without stimulation. But getting such information from dangerous wild animals is, well, dangerous, so we don't
know for sure.

Whatever the mechanism, though, the underlying puzzle is
the same. Massive stimulation is wildly extravagant. In nature,
wild extravagance quickly vanishes if it is not beneficial. If some
lionesses needed less stimulation to become pregnant—and there
were no disadvantage to having less—the levels would fall. But
they haven't. So the question stands: Why do lionesses need so
much encouragement to become pregnant?

Possibly it has something to do with the structure of lion society. Lionesses live together in family groups, known as prides. A
band of males lives with a pride, fighting off challenges from
other bands of males. If the resident band is defeated, new males
take over and kill any cubs they can find—the loss of cubs stops
the females from producing milk and brings them back into heat.
Frequent changes of males are therefore bad from the females'
point of view. So perhaps the extraordinary virility demanded of

lions is a test in which the females make sure their lions are strong and will be able to defend the pride for at least a couple of years. In keeping with this idea, there is some evidence that lionesses have lower fertility at the start of a band's tenure, as if they are testing the new men out. However, this is at best a partial explanation for the extravagance. Even when lionesses know their lions, they still copulate several hundred times whenever they come into heat.

Could female promiscuity be the reason for such extravagance? This can account for type I sex mania in some species. Among golden hamsters, for example, the more energetically a female is stimulated by one partner, the less likely she is to respond to another. Among rats, vigorous stimulation will not necessarily put a female off other partners—but if her first lover is sufficiently stimulating, he is more likely to be the father of her children. And among crested tits—small songbirds—females constantly beg for sex. A male who can't keep up with his partner's appetites will find himself cuckolded. For lions, though, the picture is ambiguous. Lions are harder to observe than hamsters or rats or crested tits, so information on lioness promiscuity is anecdotal. By some accounts, a lioness in heat will go off alone with her partner for a few days; according to others, lionesses change mates once a day. True, genetic analysis shows that it's rare for cubs in a litter not to have the same father, but that may not tell us much. If lionesses are like rats (excuse the comparison), the cubs' paternity may not reflect the mother's virtue so much as the outstanding performance of one or another of her mates.

How to resolve the matter? Since an experiment is obviously out of the question, we could compare lions with other cat species: because all cats are related, similar behaviors may have similar underlying causes. Unfortunately, the comparison deepens the mystery: although some species of cat copulate as much as lions do, they have no other obvious traits in common. For example, sex mania cannot be explained by the fact that lions live in groups. Solitary cats such as leopards and tigers also copulate like lunatics while the female is in heat. Nor is it a Big Cat Thing. Although large cats like pumas, leopards, tigers, and jaguars copulate like lions, cheetahs and snow leopards do not. Moreover, the tiny sand cat—a little-known species that preys on rodents in the deserts of the Middle East and Central Asia—copulates wildly. Other small cats, such as the bobcat and the tree ocelot, do not. And frustratingly little is known about the tendency of females in these various species to be promiscuous. For now, I'd say that female promiscuity is the best hypothesis to explain lioness behavior, but an honest jury would have to say the case is simply not proven.

I'd like to leave you with one final thought. Giant water bugs are type II sex maniacs, the male hogging the female to make sure no other man gets his hook in. The reason is that giant water bugs are devoted dads, carrying eggs about on their backs and then helping the young water bugs to hatch. Female giant water bugs don't like to mate with males already encumbered with a brood, so most males only get a chance to mate with one female at a time. They make the most of it. One male I heard of insisted on copulating more than a hundred times in thirty-six hours—or almost once per egg. Surely you're not going to be outdone by a water bug?

༚

*Dear Dr. Tatiana,*

*I think I must be deformed. I'm a long-tailed dance fly and I go to all the right parties, but night after night I'm spurned. The guys won't come near me, let alone try to seduce me with a fancy*

*dinner. I've noticed that all the other girls look like flying saucers,*
*but I look like a regular fly. Is there anything I can do?*

*Quasimoda in Delaware*

Your case is curious. In long-tailed dance-fly culture, it's traditional to mix food and sex. An hour before sunset, the male captures a suitable insect—a juicy mayfly, perhaps—and then goes to
find a female, taking his quarry with him so she can eat it while
they make love. Females congregate and wait for males to arrive.
Unlike many insects, however, long-tailed dance flies don't rendezvous at hills or tree stumps. They meet at clearings in the
forest, where the females' bodies show up in silhouette against
the sky.

Male long-tailed dance flies are fussy and prefer to give the
insect they've caught to the largest females. We don't know pre-
cisely why. In a cousin of yours, males desire large females
because they are closer to laying eggs, which means there is less
risk that the female will mate with someone else in the interim. In
your case, though, size does not indicate that egg laying is imminent. Still, in species from insects to fish, larger females are more
fecund, so your size may indicate how many eggs you are capable
of making. Be that as it may, female long-tailed dance flies have
evolved an unorthodox way of displaying how big they are. They
have two inflatable sacs, one on each side of the abdomen. Before
a female joins the party, she sits on a bush, gulping air, blowing
herself up to three or four times her normal width. Try it. I think
you'll find it'll give you the flying-saucer look.

In many species, females will only mate with males bearing
gifts. Males who can't produce the goods are rejected. Males
bearing small gifts are often punished by not being allowed to
copulate for long. This might explain why the hunting spider

*Pisaura mirabilis,* the only spider known to give presents, takes the trouble to wrap them in silk. The more silk, the more time the female takes to feed, even when the meal itself is paltry. Perhaps the fancy gift wrapping persuades the female to overlook the meagerness of the offering.

Presents take many forms, depending on the species of the giver. Often, presents are edible secretions that include proteins and other nutrients. Look at the tropical cockroach *Xestoblatta hamata.* After sex, females feast on anal secretions produced by their mates, eating right off the plate, so to speak. In many species, the secretions are not taken orally but are delivered in the same package as the sperm. During copulation in the moth *Utetheisa ornatrix,* the male passes the female a chemical that protects her from spiders. The spiders find her so disgusting that they actually free her from their webs as soon as she gets stuck, cutting away the threads that have ensnared her. In a bizarre variant, during copulation the beautiful and aptly named scarlet-bodied wasp moth covers his mate with a veil of filaments impregnated with spider repellent. Not all presents are so practical, however. Among balloon flies—which are related to dance flies—males make the female a large white silk balloon to play with while they make love.

As the gifts get more expensive, males get more careful about who receives their largesse. After all, you wouldn't take just anyone to the Ritz. Among Mormon crickets—flightless cousins of crickets and katydids—a male finds himself in severely reduced circumstances after just one mating. He secretes his gift—and loses a quarter of his mass in doing so. You can bet that these fellows are choosy and reject all but the largest mates. Many butterflies suffer a similar loss; once a male has mated, he will be unable to produce a large present quickly. Again, male butterflies are notoriously finicky about whom they bed.

With long-tailed dance flies, however, it is not so much the expense of the present that makes the guys so picky. Rather, they are because they can be. Female long-tailed dance flies have lost the ability to hunt and are thus completely dependent on males for food. You'd better puff yourself up to impress them.

∽

There's more to being a male than dropping your trousers. Good lovemaking is physically demanding—especially in species where females take additional mates. Then, sperm is anything but cheap. Not only will you have to ejaculate huge numbers of sperm, you may find yourself able to ejaculate only infrequently. This is bad news. If a female finds you inadequate, she won't wait around. She'll replace you. So before leaping into bed with the first dewy young miss who happens along, remember how Lord Chesterfield, an eighteenth-century Englishman, is said to have described sex to his son: "The pleasure is momentary, the position ridiculous, and the expense damnable."

# · 3 ·
# FRUITS OF KNOWLEDGE

---

**W**hy do females sleep around? Because, as a rule, loose females have more and healthier children. The reasons for this vary enormously, however, from species to species.

---

*Dear Dr. Tatiana,*

*My name's Vikram and I'm a bronze-winged jacana. I'm a member of a harem and I've built a nest and everything, but my female won't pay attention to me and won't give me any eggs to incubate. What am I doing wrong?*

*Neglected Househusband in Tamil Nadu*

Talk louder, shout, yell yourself hoarse. Among bronze-winged jacanas, that's about all that gets a female's attention as she skitters across the lily pads on her elegant long toes. After all, she's a busy bird. She has a big territory to defend, other males to mate with, eggs to lay. You're not going to get anywhere by hanging around like a wet feather—you have to make yourself heard.

Female jacanas have matters well in hand. They mate with all the males in their harem (often as many as four), lay a clutch for one, then repeat. Males do all the child care—so a male who is sitting on eggs or feeding chicks is excused from copulation. A female with a harem may thus have four times as many chicks as a female who has one partner. It just shows what you can do if you've got good help.

More help is one reason lots of females are promiscuous. Consider the greater rhea, a flightless South American bird that resembles the ostrich. A male receives eggs from several females. He incubates all the eggs and rears all the chicks. At the right time of year on the pampas, you can see males and their little flocks wandering about, the male whistling softly to keep them all together. The females, meanwhile, go off to mate and lay other clutches for other males to look after. Indeed, this system—where males look after the young from several females, and females spread their broods between several males—is common, especially among fish.

Even in systems where females are not so shamelessly promiscuous, they may still get more help if they sleep around. Among red-winged blackbirds—which have not red wings but red epaulettes—every male a female has bonked will come flying to her defense if her nest is attacked. This can be a delicate balance to strike: if a female's main squeeze suspects her of cheating, he may stop helping her. For example, among reed buntings, small brown songbirds, males won't make as much effort to feed the chicks if they suspect the chicks aren't theirs.

This kind of thing doesn't seem to bother bronze-winged jacanas, though. As far as we can tell—although the genetic analyses are not yet complete—males routinely rear some chicks that are not theirs. Why do they put up with it? They don't have a choice. In jacanas, females rule the roost. A female bronze-winged jacana is 60 percent heavier than a male; he has to toe the line or leave the game. How did such a system evolve? I wish I could tell you. I'm sure a lot of women out there would like to know.

∽

*Dear Dr. Tatiana,*

*I'm an orange-rumped honeyguide, and I own prime real estate: a cliff face that has several giant-honeybee nests. Lots of girls visit me, and they always let me have sex with them. They say they do it because they love me, but I worry that the real reason is that afterward they get to gorge on beeswax. Indeed, I'm starting to suspect they act the same with all the guys. Can you reassure me?*

*Waxing Suspicious in the Himalayas*

Honeypot for honeycomb? Sounds like a fair trade to me. After all, you have something they want—beeswax. And they have something you want—eggs to be fertilized. So I see no reason why the girls should give sex for free or why they should limit themselves to a single sugar daddy. Here's a thought you can comfort yourself with, though. Female orange-rumped honeyguides may not be paragons of virtue, but they are not indiscriminate birds—oh no. They only mate with males who hold territories—and who thus have beeswax to offer—so that rules out a lot of fellows.

Boys, I'm afraid the way to a woman's bed is often through her stomach. Remember the long-tailed dance fly? Well, females in many other species insist on trading sex for food. There are several ways you can meet a girl's appetites. First, like our orange-rumped friend, you can defend a territory that contains food that females need, and charge females at least one copulation per admission to the site. Second, you can hunt for food to give the female. Third, you can secrete gifts using food that you have eaten—although I wouldn't recommend this unless you are an insect.

Unfortunately, however, while food can buy you sex, it can't buy you love. In species where females swap sex for food, loose females typically eat better—and have more offspring. Take the common field grasshopper. Given a choice, females will exchange nourishment for sex with as many as twenty-five different males. Promiscuous females lay more batches of eggs—and more eggs per batch—than females who only get to mate once. Particularly when times are hard and food is scarce, this easily makes up for the hour and a half it takes to have grasshopping sex. In the green-veined white butterfly, a critter fond of damp meadows, a virgin male ejaculates a sperm packet roughly 15 percent of his weight. As well as sperm, the packet contains nutritious substances, and females who have sex with several virgins lay more and bigger eggs, and live longer, than females who do it with only one. But if their lovers are not virgins, females are even more inclined to promiscuity. Males who've already mated cannot come up with the goods: their sperm packet is a mere half the size of their virgin glory. Females compensate for the loss of nutrients by copulating with other fellows all the more.

Life's tough. Girls, don't forget that boys go to tremendous effort to find or produce food—and yet, your screwing a fellow doesn't even guarantee that you'll use his sperm. This observation has generated a vigorous controversy about why boys bother to feed girls at all. Some people argue that a good meal is just a way for a guy to get into a girl's pants. Others say that he's trying to be a good father, giving essential nutrients to the female so that she can incorporate them in her eggs—and thus give his children a good start in life. The eggs of hundreds of species have been analyzed to see whether any substances handed over by the males have indeed been incorporated. The results vary. Sometimes a female channels the nutrients to her eggs. Sometimes she nourishes herself. And sometimes nutrients given by one male go into eggs sired by another. But this shouldn't surprise you. A girl is always going to want to copulate with the males who offer the most; what she does with the offerings is up to her. All a guy can do is hope he satisfied her. Here's hoping.

∽

*Dear Dr. Tatiana,*

*I'm a male shiner perch, and I've heard that loose women are taking over my species because they have more children than anyone else. How can I stop this dreadful fraying of the moral fiber of tomorrow's perches?*

*Outraged in Baja*

It is true that in many species the females who mate the most have more and healthier children than girls who restrain themselves. But you can relax: this doesn't seem to be true in yours. Yes, female shiner perches are forward fishes, chasing males through the surf rather than waiting to be chased. But even the big females, who are likely to seduce several males, don't have more, bigger, or healthier embryos than the females who mate only once.

So why do female shiners chase guys? No one knows for sure. One thought is that they are hedging their bets against the possible sterility of their mates. In general, I don't think this is a convincing explanation for promiscuous behavior: in most species, females mate much more than is necessary to avoid sterility. Nor do I accept the popular notion that females who store sperm only remate because they've run out of it. To be sure, females can run out of sperm. In most species, though, they leap back into bed long before sperm depletion becomes a problem. And while I'm dismissing ideas, let me also put to rest the idea that females are promiscuous in order to increase the genetic diversity of their offspring. Genetic diversity is always a consequence of female promiscuity but rarely a cause.

To get back to shiner perch, however, the theory that promiscuous females are hedging against male sterility is plausible. Shiner perch are unusual fish. Rather than shed eggs and sperm into the sea as so many fish do, they copulate, the male placing sperm packets into the female's genital orifice. What's more, the females don't lay eggs—they actually give birth. But there's a long delay—several months—between the time females mate and the time they fertilize their eggs. By then, males have lost interest in sex and their testes have shriveled for the winter. If a female had only one mate and he turned out to be sterile, she would lose out on breeding for the whole year. Surely you wouldn't want that?

You do, though, raise an interesting matter. If a proclivity for promiscuity is genetic, then yes, promiscuous behavior will become more common if loose females tend to have more children than their monogamous peers. That's how natural selection works. Unfortunately, though, we don't know much about the genes that influence female sexual behavior in any animal, let alone in shiner perch. What I can say is that in both the fruit fly *Drosophila melanogaster* and the field cricket *Gryllus integer*, females do differ in their inclination to have fun, and this can largely be accounted for by genetic differences. So, boys, take a long look at your girlfriend's mother—if she was a good-time girl, the odds are on that your girlfriend will be too.

෨

*Dear Dr. Tatiana,*

*I'm consumed with existential angst. I'm a stalk-eyed fly,* Cyrtodiopsis dalmanni, *and every night girls line up to mate with me, but I rarely see the same girls twice. Worse, few of my visitors are virgins—and I know that they'll go on to other guys night after night after night. What are they looking for, and why can't I satisfy them?*

*Feeling Inadequate in Malaysia*

True, female stalk-eyed flies are massively promiscuous. But they have exacting taste. Females of your species are irresistibly drawn to a fellow with stalks so imperial that his eyes are set farther apart than his body is long. When dusk falls in the tropics, as you know from personal experience, stalk-eyed flies gather at streams, where they settle for the night on the fine root hairs that stick out under the banks. Females head for the root hairs presided over by the male with the longest stalks—and in the morning, they copulate with him before flying off for another day's foraging.

You see, girls may *say* they want a kind, sensitive, devoted guy—that personality matters more than looks—but the truth is, in many species, females are body fascists. That's why as a rule it's the males that have ridiculously long tails or fancy head-dresses—or eyes perched on the ends of long, stiff stalks.

Darwin was perplexed by the extravagant ornaments and decorations that grace the features of so many males. The evolution of weapons was easy to explain. Not so ornaments. All too often, ornaments seem to work against natural selection by making it much harder to survive. The peacock's tail might look superb—but have you ever seen peacocks fly? It's a ludicrous sight. They lumber through the air, an easy snack for a tiger.

To explain how huge tails and other frivolities could evolve despite the obviously greater risk of being eaten, Darwin came up with an idea even more radical than evolution by natural selection. He called it "sexual selection." That is, males have these ornaments because females prefer to mate with the most beautiful males. Since beautiful males would therefore go on to have more children, enhanced sex appeal could more than make up for the risks involved in having a huge tail.

Darwin was ridiculed. Females have a sense of beauty? Ludicrous. Females the ones who choose? Absurd. But as usual, he was right. These days, no one would dispute that females in many species actively decide whom to have sex with or that their preferences can drive the evolution of extravagant ornaments. What is still fiercely argued, though, is what they get by choosing the longest eyestalks or the biggest tail. Is it just sex appeal? Or is it something more?

In theory, it could be either. Sheer sex appeal can evolve through something known as "Fisher's runaway process," or the advantage of having "sexy sons." Who's Fisher? Ronald Fisher was one of the great mathematical geneticists of the twentieth century. According to his model, female taste is initially arbitrary: girls like long tails, say, just because they do. But because of this, males with the longest tails mate the most. Females who prefer the males with the longest tails have sons with the longest tails (the sexy sons)—who also mate the most. And so on. The result? Tails get longer and longer. When does it stop? One way it could stop is when the disadvantage of having a slightly longer tail—like becoming a predator's lunch—outweighs the advantage of being just that little bit sexier.

Alternatively, beauty might be more than skin deep. According to the "good genes" hypothesis, ridiculously long tails, fancy headdresses, or absurdly long eyestalks tell a girl not only that a boy has genes for long tails, fancy headdresses, and so on but that he has good genes in general. In other words, males with the longest tails are simply the best: they're in the best shape because they have the best genes—and their offspring will be more likely to survive.

Both the runaway process and the good genes theory make sense in principle. As usual, though, it's difficult to work out what Mother Nature's been up to in any particular case. Let me give you a couple of examples.

When biologists study wild birds, they often put little colored bracelets on the birds' legs so that they can tell who's who from far away. Evidently, bracelets are not a characteristic that can be inherited. Yet it turns out that female zebra finches find a male sporting red bracelets to be extremely sexy. So sexy, indeed, that they will lay him an extra clutch of eggs. Green bracelets, however, have no such charm. Perhaps females find that green clashes with a fellow's orange legs; in any case, males wearing green bracelets have no luck with the girls.

But although this preference shows females can pick a mate based on sex appeal alone—the basic requirement for a runaway process—in more natural situations it is often exceedingly hard to tell whether females are mating because they find the males sexy or because they're hoping for good genes. An example of the difficulties comes from peacocks. Peachicks sired by males with fancier tails survive at a higher rate than those fathered by more humdrum-looking males—which at first glance supports the good genes hypothesis. The trouble is, there's always the possibility that the effect appears simply because females who mate with sexy males take more trouble over their offspring. Among mallards, a duck who mates with the sexiest drake lays larger eggs than she does for an ugly fellow. Big eggs are a crucial factor in a chick's early survival. But it turns out that a drake's genes have no bearing on the size of the eggs. The difference in bigness can be explained entirely by the mother's tender loving care. So why does she bother? Who knows. Perhaps it's Fisher coming through the back door: perhaps ducks with sexy mates put more effort into their offspring because they know their ducklings will grow up to be sexy too. Is something like this also going on in peacocks? We don't know. Although peahens don't alter the size of their eggs, they could be making more subtle adjustments. Female zebra finches mated to attractive males, for example, increase the testosterone composition of their eggs—thus accelerating the speed at which the chicks grow.

Your situation is similarly complex, my stalky friend. When females mate with you, are they after good genes or sexy sons? There is evidence that if the larval environment is harsh, only a few males have the genes to grow great stalks. Again, at first glance, this supports the good genes idea: females select the males most able to cope with a tough environment. Alas, without information on whether these males have offspring who are more likely to survive than the offspring of other males, we cannot arbitrate. In picking males with the longest eyestalks, the females also increase the odds that their sons will be sexy too.

∽

*Dear Dr. Tatiana,*

*I'm a harlequin-beetle-riding pseudoscorpion. At least, I should be. But when I found a beetle to ride, I wasn't allowed on board. Some big thug of a pseudoscorpion helped my girlfriend on but pushed me away as the beetle took off. She went with him happily, and I just know she's having sex with him, the tramp. Meanwhile, here I am marooned, a stuck-on-a-rotting-log pseudoscorpion. I've tried waving my pincers at harlequin beetles flying overhead, but none has landed. How can I get off this log and find a girl who will be true?*

*Stranded in Panama*

Boy, have you got a bundle of problems. Let's start with the most pressing one—getting you off that log. You won't get anywhere by waving your pincers, I'm afraid. Here's the scoop. Harlequin-beetle-riding pseudoscorpions live on rotting logs—the fallen branches of fig trees, preferably. The only problem with this otherwise excellent arrangement is that sooner or later a given log will decay completely; when it does, anyone in residence will perish. So, how can you escape from the log before it has rotted to nothing? Enter the harlequin beetle.

Harlequin beetles are magnificent, their jet-black wing covers decorated with jagged red stripes. More germane than their looks, however, is where they make their homes—in rotting logs. The cycle begins when a female harlequin beetle lays her eggs in a freshly fallen fig tree. Her children develop in the wood; after several months, they emerge fully grown. This is your moment. Pseudoscorpions are tiny—far smaller than proper scorpions (although the real distinction is that you have no sting). This means that you can stow away under a harlequin beetle's wing covers and fly off to a new home when the harlequin beetle flies to fresh logs to find mates and lay eggs.

But as you found out, space under the wings is limited. Even on a large beetle, no more than about thirty pseudoscorpions can clamber on board. Worse, a big male pseudoscorpion can easily defend this space from rivals. With bogus gallantry, he lets females on and keeps other males off. Then, when they take to the skies, he will have sex with as many of his companions as possible. Right now, just as you feared, your girlfriend is probably squatting over a packet of sperm that the "big thug" has deposited for her on the harlequin beetle's back. Sorry about that. And I'm afraid I have some other bad news. Harlequin beetles already out in the world are not attracted to old logs, so there will be no casual passersby. If you can't find and board another beetle as it emerges, you'll be marooned forever, a helpless member of a doomed population.

But you can console yourself a little. When the harlequin beetle carrying your girlfriend arrives at a new log, she and the other females will disembark. Your girlfriend will give birth to her family there in the decaying wood. Because she has the ability to store sperm, some of her children will probably be yours. So as you sit on your island, think about all the baby pseudoscorpions bearing your likeness.

As to finding a female harlequin-beetle-riding pseudoscorpion who will be true—well, you might as well try to catch a falling star. These females like to gallivant: they regularly spurn the attentions of previous lovers, preferring to make fresh conquests. Why? It's simple. Females who mate with two different males are more likely to have children than females who mate with the same male twice. This is not because lots of males are sterile. Rather, females who mate with only one male are more likely to abort their broods. Apparently genes from the father and genes from the mother are sometimes incompatible and cannot act in concert to produce children. By mating with several different males, females avoid this problem.

My hunch is that avoiding genetic incompatibilities will turn out to be the reason that females of many species are promiscuous. This is still a new idea and has been tested only in obvious candidates such as the honeybee. Incompatibilities between male and female genes, however, are probably common. In the beetle *Callosobruchus maculatus*, for example, whether or not a male is successful in sperm competition depends in part on the genes of his mate. And certainly, genetic incompatibility is an important cause of infertility in a number of species.

In humans, for example, perhaps 10 percent of couples are infertile. Of those, between 10 and 20 percent of cases are apparently due not to either partner's being sterile but to a genetic incompatibility. Furthermore, some women are prone to the spontaneous abortion of healthy fetuses. Again, the problem often lies with interactions between the partners' genes. Does this cause women to be unfaithful? I can only wonder.

Still, there is tantalizing tentative evidence that people might have inbuilt mechanisms to avoid such problems in the first place. Consider this. The genes typically implicated in spontaneous abortions are part of a vast gene complex known as the major histocompatibility complex, or MHC. These genes are important in the immune system. They help determine resistance to infectious diseases and, in humans, cause the rejection of transplanted organs. There may be as many as a thousand different genes in the complex—and some of them come in as many as a thousand different forms, so they are exceptionally variable. Indeed, the MHC seems to give each individual a unique odor. For example, people and mice can both smell the difference between mice that are genetically identical except at the MHC. And in a number of "smelly T-shirt" experiments—experiments in which humans sniff T-shirts that have been worn for a couple of days by members of the opposite sex—people consistently prefer the smells of those whose genes at this complex are different from their own. Yes, you guessed it. Spontaneous abortions are more likely when couples match at particular MHC genes.

Of course, a woman can't love on smell alone, and so when you look at who actually pairs up with whom, you don't find a consistent pattern. But it is intriguing that in the smelly T-shirt experiments, the only women who liked the smells of men who matched them at the MHC happened to be those taking the oral contraceptive pill. We don't know why. But I'm sure you'll agree that if this result turns out to be generally true, the implications would be most disturbing.

∽

*Dear Dr. Tatiana,*

*I'd prefer to keep my identity secret, since I am writing to you not about me or my species but about my noisy neighbors—a group of chimpanzees. When those girls come into heat, it's enough to make a harlot blush. Yesterday I saw a girl screw eight different fellows in fifteen minutes. Another time I saw one swing between seven fellows, going at it eighty-four times in eight days. Why are they such sluts?*

*Mind Boggling and Eyes Popping in the Ivory Coast*

You raise an excellent question. The extraordinary promiscuity of female chimpanzees has intrigued many a scientist, and to be frank, we don't know why they are so incredibly wild. However, two theories are regularly bandied about.

The first is that female chimpanzees mate promiscuously to create competition between sperm from different males. In other words, sperm competition is not merely the consequence of females mating with more than one male, but the cause. I know this sounds outlandish. But it gets wheeled out to excuse the licentious behavior of females in many species, so it's an idea worth scrutinizing. Here's how it's supposed to work.

The starting assumption is that some males are much better at fertilizing eggs than others. The reason they are better doesn't matter much—what matters is that the ability is heritable. That is, excellence at fertilizing eggs must have a genetic basis and those genes must be handed on from father to son. Then females who sleep around—and thereby encourage sperm competition— will have sons who are better at fertilizing eggs than females who mate once.

The evidence, however, is circumstantial at best. I don't deny that it's possible, but I have yet to see a rigorous demonstration that setting up a sperm race is the main reason that females of any species sleep around. Although biologists have arranged endless contests to find out who wins when sperm are in competition, a huge number of variables affect the outcome; there is no general rule. Sometimes it's a question of who's on first, sometimes it's a matter of timing, sometimes it depends on the number of males competing—and so on. Certainly, many of the variables are not under genetic control. In the rat, for example, the female's reproductive tract is bifurcated—and the outcome of sperm competition often differs between the left and right halves.

But suppose you succeed in showing that one guy always beats everyone else. That does not mean his superiority is passed to his sons. At least one crucial ingredient for successful sperm cannot be passed from father to son, namely, the engines that sperm need in order to move. These engines are known as mitochondria. They are tiny organs that generate energy for cells. In most animals, mitochondria are inherited only from the mother. Can engine trouble ruin a guy's chances at fertilizing eggs? You bet. Faulty mitochondria cause infertility in men, rams, and roosters, for example. Conversely, some guys with otherwise unremarkable sperm may find theirs move as if turbocharged, as if someone put rocket engines on a wheelbarrow. So you see the difficulty. Just demonstrating that one guy's sperm is consistently more competitive is not enough. You have to show that this is due to a characteristic that can be inherited. I'd even guess that unreliable engines could explain why sperm tend to be larger, more numerous, and more complicated in species where there is sperm competition. All those other traits are heritable and may partially compensate for unreliable engines.

The second theory that aims to explain why female chimpanzees are so promiscuous is the obfuscation theory. The idea here is that by mating with every guy in sight females can create confusion over the paternity of their child. And clearly, if a girl's enjoyed the gang bangs you've described, even she won't know who's the dad. Why would this be an advantage? Well, perhaps if a male thinks a child may be his, he will refrain from killing it. Infanticide is a risk, after all: male chimpanzees do sometimes murder infants. But whether they are more likely to murder the children of females they haven't copulated with is, for now, a mystery.

As for you, if your biology allows for it, might I suggest a move to a nicer neighborhood?

ᝪ

*Dear Dr. Tatiana,*

*I'm a yellow dung fly, and I've heard rumors that in my species sperm are actually chosen by the egg. Is this true, and if so, what can I do to make my sperm more attractive?*

*The Dandy on the Cowpat*

That's a tricky one. The question of whether eggs—or females— actively select one sperm over another is contentious. To be sure, females can reject sperm from particular males. Consider the Caribbean reef squid. Males place packets of sperm anywhere on the female's head or tentacles. The female either moves the sperm packet to her sperm storage organ, the seminal receptacle, or she picks it off and throws it away. Then there's the farmyard chicken. Females who copulate with a male low in the pecking order are likely to eject his sperm as he dismounts. But whether females store sperm from several males and then choose the winning sperm or whether each egg actively prefers particular sperm, that's another matter altogether. I know of only one case where something like this does seem to be going on.

Have you ever met a comb jelly, or ctenophore? No? There are about a hundred known species, although given their penchant for life in the vasty deep, many more probably await discovery. To the untutored eye, comb jellies resemble jellyfish—the typical member of each group is translucent, lives in open water, and can give a nasty sting. The similarity, however, is superficial. For one thing, comb jellies have firmer bodies. But the chief distinction is the eponymous "comb"—eight ridges of cilia that run down the sides of each comb jelly and that wave in unison to propel the animal languidly along. *Beroë ovata*, one of the biggest species of comb jelly, is shaped like a bell. For whom does it toll? Ask not. If you're another comb jelly, I'm afraid it tolls for thee: *Beroë* is a voracious predator, swimming along mouth first to engulf comb jellies of other species. (If a *Beroë*'s not hungry, it zips its mouth shut to cut down on drag.) But to get to the point, *Beroë* has some singular reproductive habits.

Like most other comb jellies, *Beroë ovata* is a hermaphrodite and sends both eggs and sperm out into the sea. Self-fertilization is rare: sperm released by the same comb jelly that released the egg are not usually allowed through the egg's outer covering. Nothing odd so far. The going only gets peculiar after the egg has been fertilized. As long as just one sperm enters the egg, the baby comb jelly will begin its development as you'd expect. But if several sperm enter the egg, things become interesting.

If several sperm penetrate a human egg, it won't develop. But for many animals—sharks, for example—polyspermy is not a terminal condition but the norm. In our friend *Beroë*, it seems to provide an arena for the ultimate in mate choice. The nucleus of the egg moves around and "visits" each of the sperm nuclei in turn, before eventually "deciding" which one to fuse with. The process can take hours—and the egg nucleus won't necessarily fuse with the last sperm it inspected but will sometimes turn around and go back to one that took its fancy earlier on. How does it decide? This system has been studied so little that it is hard even to speculate.

Of course, it's possible that, unbeknownst to us, similar shenanigans go on in other species. Finding out will be difficult, however. You see, *Beroë* eggs are fertilized outside the body and are easy to look at under a microscope. In species with internal fertilization—such as those that copulate—it's hard to get a microscope to the scene of the action. That means we can only draw inferences about sperm selection. Just because one male succeeds in fertilizing more of a female's eggs than another doesn't mean the sperm have been expressly chosen. The successful sperm may be more competitive or more compatible. Or the effect may be due to chance. In the mallard, for example, females who have been artificially inseminated with a mixture of sperm from several males tend to use the sperm of one male for a given clutch. Which male is the lucky one changes each time, however, even though the female receives the same mixture of sperm—suggesting that the effect is due to sperm clumping rather than to an active preference for the sperm of one male in particular.

As for yellow dung flies, claims have been made that the female's decision to use one male's sperm rather than another's depends on whether she lays her eggs on a cowpat in the shade or one in the sun. It is a fascinating idea but extraordinary claims require extraordinary evidence, which we do not currently have. In any event, if I were you I'd concentrate on eliminating the option of your mate's choosing sperm. If a male yellow dung fly copulates for long enough, he can displace the sperm of previous males (to achieve this effect, small males have to copulate for longer than big males, because small males transfer sperm more slowly). Having replaced the sperm of your predecessors with your own, you should then guard the female until she has laid her eggs. That way, you won't have to worry: your sperm will be the only ones available. Go for it!

∽

So you see, there are lots of reasons females might play the field, although we don't necessarily know the reasons in any given instance. Just in case you meet a girl on the prowl and you want to understand her motives, here's a checklist of possibilities:

❑ She has run out of sperm
❑ Her other lovers were sterile

❏ Her other lovers had lousy genes
❏ Her other lovers had incompatible genes
❏ Her other lovers were ugly
❏ She wants diversity in her children
❏ She wants you for your food
❏ She wants help raising her kids
❏ She wants to enter your sperm in a competition
❏ She wants to give herself or her eggs a selection of sperm
  to choose from
❏ She wants to confuse everyone about who's the father

You'll notice that one obvious possibility is missing from the
list—namely, that females sleep around for pleasure. The omis-
sion is deliberate: we know next to nothing about the evolution of
sexual pleasure. I'd bet, though, that sexual pleasure is most
likely to evolve when females have a lot to gain from promiscuous
behavior.

Folks, it's time to bury forever the notion that female promiscuity is an unfortunate accident—a "malfunction," the result of
coercion, or simply a last resort to get a pesky guy to go away
(known as "convenience polyandry," this notion presumes that a
male will stop harassing a female once he's had his way with her).
Which is not to say that females are never coerced or harassed
into having sex. Or that sleeping around is always good. In the
wasp *Macrocentrus ancylivorus*, for example, a female who mates
too often gets clogged up with sperm and can't fertilize her eggs.
But like it or not, in countless species—from grasshoppers to
fruit flies, pseudoscorpions to spiders, red-winged blackbirds to
prairie dogs—it is not simply that females mate with lots of
males. It's that doing so is *good* for them: promiscuous females
have more and healthier children. Natural selection, it seems,
often smiles on strumpets. Sorry, boys.

# ·4·
# SWORDS OR PISTOLS

The art of dueling is knowing when to fight, when to flee—and when to play dirty.

*Dear Dr. Tatiana,*

*I'm a fig wasp, and I'm in a panic. All the males I know are psychos. Instead of wooing us girls, they bite each other in half. What can I do to stop them?*

*Give Peace a Chance in Ribeirão Prêto*

You can't do anything, I'm afraid. Your society is one of the most violent on earth. For every crop of figs from your fig tree, millions of young male wasps have died in combat. That's why they all have huge heads, gigantic scything mandibles, and heavily armored shoulders. And that's why they all seem deranged: in your species, it's kill or be killed. Still, you shouldn't fret. Once he's vanquished his rivals, the winner will mate with you. Why has such brutality evolved? The answer lies in your unusual lifestyle.

Throughout the sun-drenched tropics, monkeys, birds, rodents, and bats feast on the fruits of the fig tree. The trees are one of nature's successes: ancient, and abundant, they come in hundreds of different species of every shape and size. Some, such as the banyan tree, spread sideways, dropping thick roots from their branches. Others are tiny shrubs. Still others are parasites that eventually strangle their host plants. But whatever their way of life, fig trees have one feature in common. All of them depend on tiny wasps to pollinate them. Without the wasps, they cannot reproduce. On the Hawaiian island of Kauai, for example, fig trees were introduced without wasps and thus have not been able to multiply.

Although each species of fig has its own private species of wasp, the different systems work in much the same way. The cycle begins when a female arrives at a fig flower, an enclosed urn with several hundred tiny flowers inside. This whole structure will later develop into a fruit. The arriving wasp struggles into the urn, often losing her wings and antennae in the process, and pollinates individual florets within. Depending on the species of wasp, she may pollinate by accident, brushing against the florets as she moves around inside the fig; or she may deliberately smear pollen on the appropriate organs of chosen flowers. After this, she lays her eggs, placing each into an ovary of a floret. Then she dies.

Her children awaken to find themselves inside fig seeds—so they are minute. Each seed is only one or two millimeters long. The young wasps eat their way out—levying a pollination tax on the fig of one seed per pollinator. This may not sound like much, but it adds up. Some trees lose more than half their seeds to the tax collector. Males emerge first and help the females out of their seeds. They mate at once, then the males gallantly chew an exit out of the fruit so that the females can escape. Since the males have no wings, they die in the fig they were born in. The females, meanwhile, having collected pollen, fly off to find a new fig to struggle into—and so on. (Does this mean that humans eat dead wasps every time they eat a fig? Yes and no. Not all cultivated fig varieties require wasps to make fruit. However, some do—and then, yes, eating figs means eating dead wasps. But it's no big deal. As I said, the wasps are tiny. And they're not poisonous. On the contrary, they provide a little extra protein.)

You'll have noticed I've said nothing so far about fighting male wasps. That's because pollinating wasps are generally peaceable types. They are not, however, the only occupants of a fig. As well as having a pollinator species, each fig species has to put up with parasitic wasps—sometimes as many as twenty-five different species. A few are pollinators that have gone bad. They live off the fig, but the pockets they once, long ago, filled with pollen stay empty. Others simply prey on the pollinators. For the most part the biology of these parasites is poorly understood, but we do know that they are often prone to violence.

The difference stems from the way that female parasitic wasps lay their eggs. Rather than crawling into the urn, most of the parasites lay their eggs by drilling through the outside. This means that these females do not have to lay all their eggs in one place as pollinators do but instead can lay a few eggs in many. At the right time of year, you can see speckles on the fig skins from the drilling of different females.

Spreading a brood among several figs will have one of two consequences. In species where population densities are low, males risk finding themselves alone in a fig. So it is not surprising that where this is likely, males have wings so that they can fly off to search for lovers outside the fig of their birth. But in species—and yours is one—where figs tend to be crowded, males do not fly. Any given fig will probably also be home to mates. The trouble is, of course, it will probably also be home to rivals. Hence the slaughter.

Mortal combat is an effective but risky way of eliminating rivals. Males are not usually interested in fighting to the death unless they have a lot to gain—and little to lose—by doing so. Dying, after all, precludes further reproduction. Thus, lethal fighting is most likely to occur in species like yours that live for only one breeding season. One-shot breeding does not by itself, however, produce routine violence. Two other factors are crucial. First, receptive females must be clumped in space and time such that a fellow's only chance to mate is here, now. Second, fighting must increase the number of females he can mate with: there is no point wasting time fighting if in doing so he is missing out on sex. For example, if females are abundant but they mate only once, then males who copulate will do better than males who fight—a dynamic thought to explain why fighting is virtually unheard of among pollinator wasps.

Only a handful of other creatures have a reputation for extreme violence, and what little we know of them is consistent with this scenario. Take the "annual" fishes of Africa and South America. Their lifestyle is almost magical. They live in puddles, ponds, and ditches that dry up for part of the year. When the puddles dry up, they die. Only their eggs survive, buried under the dried mud, waiting for the next rains. Collect mud, add water—and presto, you get fish! You can see why people believed in spontaneous generation. When the rains come, time is short. As there is no chance to move to a new neighborhood, everyone tries to be the big fish in the small pond: the males of some of these species are among the most pugnacious fishes known. If I had to bet, I'd predict that the more ephemeral the puddle, the fiercer the fighting.

Or take gladiator frogs. These brown tropical frogs have evolved switchblades: on each hand, just above the thumb, males have a sharp, retractable spine that is shaped like a scythe. Most of the time, they keep it sheathed in folds of flesh. But when two male frogs fight, they rake these spines across each other's faces, aiming for the eyes and eardrums of their opponent. Although we don't know the death toll in the wild, we do know fights can be lethal. As you would expect, competition for mates is fierce. And also consistent with the hypothesis, gladiator frogs have short lives. Even without the fighting, few survive from one breeding season to the next.

Perhaps the strangest example of violence is not from an animal but from a plant, the orchid *Catasetum ochraceum*, and its relations. In these species, female flowers receive pollen from only one pollinator. Immediately after pollination, they swell shut and set about making fruit. Competition between male flowers to be "the one" is therefore fierce. But since they are plants, they can hardly fight it out man to man. Instead, they direct their aggression at a hapless intermediary—a bee. The male flowers assault any bee who ventures inside them, throwing sticky sacks of pollen onto the bees' backs. In one species, the flower throws the pollen at the staggering velocity of 323 centimeters per second. Since the pollen sacks are also large—they can be as much as 23 percent of a bee's weight—the bees don't like this treatment. After one such attack, they avoid all other males, visiting only the gentle female flowers.

But for brutality, nothing rivals wasps in a fig. One scientist I know tried to study fighting in fig wasps—only to find that he was never there in time. Whenever he slit open a fig, there would be just one male left alive, celebrating his victories by mating with all the females in the fruit. So as you see, there's a reason all the guys you know have such terrifying demeanors—and maybe now that you understand, you can forgive them for being such maniacs.

∽

*Dear Dr. Tatiana,*

*Perhaps you can help. I don't know what's happened to me. I'm a twenty-seven-year-old African elephant, and I used to enjoy showering at the water hole and other idle pleasures. But the joy has gone from life. I feel angry all the time—if I see another bull elephant, I want to kill him. And I'm obsessed with sex. Night after night I have erotic dreams, and the sight of a beautiful cow sends me into a frenzy. Worst of all, my penis has turned green. Am I ill?*

*Anxious in Amboseli*

Uncontrollable aggression, obsessive lust, morbid anxiety about your sexual health: this all sounds normal for a fellow in his late twenties. It's nothing to worry about. You've just got a case of SINBAD: Single Income, No Babe, Absolutely Desperate. Unfortunately for you, however, you're likely to be in this state for much of the next twenty years. Female elephants prefer older males. Until you're bigger, the cows will run away from you—and their mothers and sisters will bellow for older bulls in the area to come and send you packing.

I'm afraid that females in many species often provoke males into fighting over them. When the mood strikes, they make themselves conspicuous—then stand back and watch the battle, before mating with the winner. Female northern elephant seals, for example, create a rumpus whenever a male tries to mount them; this has the immediate effect of summoning every other fellow on the beach, even waking those who've been snoozing. The Burmese jungle fowl—the ancestor of the farmyard chicken—gives a loud squawk after laying an egg, an odd thing to do given that it immediately tells hungry predators there are eggs on the menu. But it also, apparently, goads any roosters in the area into fisticuffs over the chance to fertilize her next egg. In my opinion, though, the worst offenders are found in *Zootermopsis nevadensis*, a termite that lives in rotting wood. In this species, males and females usually live together as couples. They meet each other on a suitable log and found a nest. During this initial period, if a male finds he doesn't like his mate, he'll probably leave. But a female will typically invite a new male into the nest, which almost always results in the two males dueling. Between bouts, the female will groom first one and then the other. (On rare occasions, a male will invite a new female into the nest—and will then encourage the girls to fight.) Finally, although female cheetahs do not, as far as I know, goad their admirers into fighting, they do find fights arousing and come into heat shortly after watching one.

Female provocation does not usually lead to a pileup of corpses, however. As the saying goes, "He who fights and runs away lives to mate another day." There's no point in fighting if you know you're going to lose—especially if by bowing out, you may be able to find another mate elsewhere. That's why even in one-shot breeders, fighting to the death is rare. If a male arrives to discover that another guy got there first, there may be some argy-bargy, but it's more likely to be a show of strength than all-out war. Take two-spotted spider mites. These tiny creatures are agricultural pests: they feed on plant cells, stabbing each cell with their mouthparts and drinking it dry. Although they are mites, they have the ability to make silk just like their spider cousins. In this species, males hunt for adolescent females—those about to undergo their final molt and become adults—and stand guard, so that they can be first come, first served. A guarding male sits on the female, with his legs draped over her body. If a second male arrives and refuses to leave when threatened, some sort of fight will break out. The two males will wave their front legs and grapple with each other, often trying to trip their opponent by attaching strands of silk to his legs. Fights can end in death, but that's rare. Usually, the smaller male retreats before it comes to that.

Size is crucial in hand-to-hand fighting. In animals from boa constrictors to humans, the larger combatant typically enjoys a large advantage, so smaller males usually back off. As a rule, savage fighting breaks out only when both contestants figure they can win—that is, when they are about the same size. Consequently, many animals have evolved bizarre rituals to assess their opponents. Remember those flies with eyes perched on the ends of long, stiff stalks? Males measure their strength by going head-to-head to compare eye spans. If there isn't a clear difference, they'll fight; otherwise, the male with the smaller eyespan leaves. Thus, you can tell when fighting is important in determining who gets to mate: males will typically have evolved to be big. Male elephants, for example, have evolved to grow throughout their lives rather than stopping after puberty like most mammals. Whereas most mammals cannot grow beyond puberty because their bones fuse together, a male elephant's bones don't fuse until he is middle-aged. That's how male elephants can grow to be more than twice the size of females.

Among elephants, though, it's not only size that matters. Bull elephants, even the largest and oldest of them, feel like fighting only at certain times of the year, when they are in the grip of a fury known as musth. For young males, a bout of musth lasts only a few days at a time, but for the oldest bulls it can continue for as long as four months. When a bull is in musth, the amount of testosterone in his blood soars to levels fifty times higher than usual. Unsurprisingly, this has profound effects on a fellow's behavior.

Males in musth display all the symptoms you complain of, and more. They wave their ears and shake their heads and constantly dribble strong-smelling urine—this is what causes the penis to develop the unfortunate greenish tinge you mentioned. Musth even affects their conversation. Most elephant conversations take place in the infrasound, well below the range of human ears but audible to other elephants for miles around. Male elephants are usually the strong, silent type—probably because, unlike the females, they haven't got large vocabularies. (In fact, it's not just that male elephants have smaller vocabularies but that male and female elephants have vocabularies that are almost entirely mutually exclusive. Elephant girls and boys couldn't discuss the same subjects even if they wanted to. I know the feeling.) But an elephant in musth rumbles constantly, announcing his desperate lust and aggressive anger to any elephant in earshot. Males in musth are also much more likely to pick fights with one another than with other males, and if the males are the same size, fights are more likely to escalate. That's bad news. A big fight can last for hours in the heat of the African sun and can certainly be fatal. In between rounds, fighting elephants uproot trees and throw logs, and if, at the end of all this, both contestants are still stand-ing, the fight finishes with the winner chasing the loser for sev-eral miles. This is probably why large males in musth often try to avoid one another—and why even large males who are not in musth stay well away from any male who is.

&#8766;

*Dear Dr. Tatiana,*

*My name's Rob, and I'm a bedbug,* Xylocoris maculipennis. *I've read that if I have sex with my friend Fergus, he'll deliver my sperm when he next has sex with Samantha. Is this for real?*

*Making Mischief between the Sheets*

First, you're not a real bedbug but a pirate bug, a close cousin. No use pretending. As to your question, it sounds to me like you've been reading too much naughty French literature (although to be fair, I've read the same stuff myself). The claim is that because you have a penis like a hypodermic needle—and because in species like yours, you jab it through the body wall—sperm injected into another male will migrate through his body and arrive in his gonads. Exciting though this sounds, it has always struck me as one of those facts journalists call "too good to check." The short answer is yes, it's possible—but it's unlikely.

I have two reasons, one practical and one theoretical, for being skeptical. On a practical level, the original claim is flimsy—based on poor data—and the experiments have never been repeated on your species. Moreover, experiments on a related species, the European bedbug, have found no evidence for males injecting each other. Theoretically, the situation you describe would be highly unstable: any male who could resist acting as a proxy would have a big advantage over one who could not, and genes for resistance should swiftly spread through the population. Which is exactly what has happened in the sea squirt *Botryllus schlosseri.*

This creature undergoes cycles of both sexual and asexual reproduction. It lives in colonies on rocks and reefs. A sea squirt looks like a tiny barrel with two siphons on the top; in a sea squirt colony, the barrels are embedded in a gel-like matrix. Looking at the adult, you'd probably never guess that these organisms are animals, let alone close relations of animals with backbones. Only the larvae, which look like simplified tadpoles, give the game away. When a larva settles on a new rock, it metamorphoses into a grown-up sea squirt and begins to reproduce asexually to form a group of genetically identical individuals that share a common blood supply.

As neighboring colonies expand, they may run into each other. At this point, they have a choice. They can either join forces and become one big colony or they can reject each other, creating a physical border, a line in the sand that neither can cross. If the colonies fuse, something sinister can happen. Cells from the individuals of one colony can travel through the common blood supply and invade the gonads of all the individuals from the other colony. It's a hostile takeover: when the time comes to reproduce sexually, members of the losing colony are forced to make eggs and sperm carrying genes that are not their own. Predictably enough, this has led to the evolution of mechanisms to avoid having your gonads hijacked. *Botryllus* colonies are highly particular about whom they fuse with: whether or not fusion goes ahead depends on a complex system that ensures close matching between genes of the two colonies. Colonies fuse only when they have similar genes—and therefore, are probably closely related. In pirate bugs and bedbugs, however, there is no evidence of measures to counter gonadal hijacking—and I bet it doesn't happen in the first place.

The males of some species do, however, have another, more plausible way of eliminating their rivals: they render them impotent. Such nastiness has been claimed for at least one spiny-headed worm, the gloriously named *Moniliformis dubius*, a scourge of cockroaches and rats. Baby worms live in the guts of cockroaches; when a cockroach is eaten by a rat, the worms grow up and have sex in the rat's intestines. Not exactly romantic, but there you go. What's important to us is that female worms mature at the same time, a situation that, as you know, promotes fierce contests. But here the contest is not fighting but sabotage by "cementing." When a male mates with a female, you see, he finishes off by capping her genitalia with a chastity belt made of a kind of cement. It turns out that males aren't shy about cementing up each other either: by applying cement to another guy's genitals, they prevent the other guy from copulating. Is it just that male spiny-headed worms can't tell the difference between males and females? Maybe. But there is some tentative evidence that when a male cements a male he doesn't transfer sperm, so perhaps the aim really is to make his rivals impotent.

Spoiling tactics have been proposed to explain mysterious goings-on in other species too. For instance, among African bat bedbugs—large bedbugs that suck the blood of roosting bats—females have a peculiar genital structure on their bellies for receiving sperm. This structure is entirely separate from where the eggs are fertilized. Curiously enough, the males have the same structure. One possibility is that they use it to sabotage rivals by fooling them into depositing sperm there, thus wasting their reproductive effort. However, since any male who could tell the difference between a male and a female would have a significant advantage, I'd be astonished if this is the real reason the males have this structure.

I suspect that such spoiling tactics are most likely to evolve when males do not pass on their sperm directly but put it into a packet—a spermatophore—that they leave lying around in the hopes that a female might pass by or that they try to persuade her to pick up. In some species, males have been reported to trample or eat any spermatophores they encounter; in others, males deposit their own packet on top, forming a kind of sperm stalagmite. And some males come between a mating couple as the sperm deposition is about to take place. The Jordan salamander, an amphibian that looks like a lizard, is an accomplished third wheel. During courtship, male and female Jordan salamanders do a dance in which he leads and she follows behind, straddling his tail until he deposits his spermatophore. Occasionally, a second male slips between the courting couple and waddles along as if he's the female. When the first male is duped into depositing his packet of sperm, the second then goes off with the bride, and the unhappy loser has to wait several days before he can make a new spermatophore. But either Jordan salamanders are fools or the risks of this happening are low enough that most of the time a male doesn't have to look over his shoulder while making love.

⌒

*Dear Dr. Tatiana,*

*It's a fiasco. I'm a hopeless male three-spined stickleback. I was watching my eggs when I heard a sudden noise. I turned to look—just for a second—and when I turned back, all my eggs had been stolen. Who would do such a terrible thing, and how can I prevent its happening again?*

*Want My Eggs Back in Vancouver*

Egg pirates—it's an old problem. All you can do is remain vigilant. The trouble is, in many fish species females prefer to lay eggs in a nest that already contains some. They take the presence of other eggs as proof that the nest is safe, that the male who owns it is particularly manly or that he's a good parent and unlikely to eat his babies. Success spawns success, you might say.

And in sticklebacks, it often spawns a black market in eggs. I don't mean that you get rough-looking fellows selling stolen eggs in the run-down parts of lakes and streams. No, the thief keeps the eggs for himself, taking them to his own nest so he can pretend he's a breed of superdad. Why are sticklebacks particularly vulnerable to egg theft? We don't know for sure. It may be that the eggs are easy to heist: unlike the eggs of most fish species, stickleback eggs stick to each other in convenient, portable clumps.

Something like this goes on in the cloud forests of Papua New Guinea and the rain forests of Australia. The aristocrats of these forests are the bowerbirds, close relations of birds of paradise. And like their relations, bowerbirds mostly eat fruit. Because they are quite big, bowerbirds are easily able to monopolize fruit trees, scattering smaller birds out of their way. Thus, like aristocrats everywhere, most of these birds have lots of free time. And so, naturally, they have a hobby. It's art.

Male bowerbirds spend weeks building and decorating elaborate "bowers." Depending on the species, the bower could be anything from a clearing strewn artfully with leaves to huts more than four meters (thirteen feet) wide or towers more than three meters (ten feet) high, woven out of sticks, painted with juice from crushed fruits, and decorated with flowers, mushrooms, feathers, snakeskins, snail shells, butterfly wings, beetle heads— or anything else that catches the artist's eye. One scientist nearly had his camera stolen by a bowerbird who wanted to add it to his decor; another almost lost his socks. Artistic styles differ greatly among populations—even populations of the same species—so that whereas flowers might be fashionable in one area, beetle wings will be all the rage in the next. Moreover, this is no random collection of junk: the objects are selected and placed with great care. If you intrude into a bower and move things around, the artist will put them back again after you've gone. If you add objects that weren't there before, he'll take them away. And if you watch an artist at work, you'll see him experimenting—trying out different items in different positions.

Why do they do this? To impress girls, of course. Females come to the bowers to mate. And one way to make your bower look even better than a rival's is to resort to theft and vandalism. Yes, I'm afraid that bowerbirds are not above foul play to further their own ends. Stealing is rife. Rare or fashionable objects vanish from one bower only to appear in another. And some bowers are regularly vandalized or completely destroyed. Vandals, like the egg pirates or any other common burglar, approach stealthily, tiptoeing through the undergrowth and freezing at the slightest sound.

Worst of all, such behavior is often rewarded. In species like yours or theirs, females don't seem to care how a male filled his nest with eggs or why he's the only one with a collection of unusual feathers. They just go for the fellow with the most lavish display. In these species, as in so many others, I'm afraid that nice guys finish last.

Boys, if you feel consumed with aggression, you're probably experiencing a testosterone surge. Keep cool. Don't plunge into battle with the first fellow you see—and above all, don't let yourself be goaded into fighting over a woman. Remember: there are few circumstances in which you should actually throw down the gauntlet and fight. If you have doubts, refer to my Rules of Engagement.

## RULES OF ENGAGEMENT

You should be prepared to fight to the death only if you can answer yes to both of the following questions:

1. Is your only chance to mate here, now?
2. Will fighting increase the number of girls you can mate with?

If the conditions are not right for fighting to the death, it may still be appropriate to trade body blows. However, you should fight only if you think you can win. Since size is the primary determinant of victory, you should always pick on someone smaller than you. Before engaging in combat, try intimidation: flex your muscles, puff up your chest, shout, do whatever it takes to show your opponent there's no point in fighting you. If, when you've done all this, you discover you're less frightening than your opponent, retreat immediately. If you find you are always running away, do not despair. I have help for you in the next chapter.

# · 5 ·

# HOW TO WIN EVEN
# IF YOU'RE A LOSER

---

What if you're poor? What if you're ugly? What if you're a wimp? What if you're a poor, ugly wimp? Relax and read on . . .

---

*Dear Dr. Tatiana,*

*I'm a sponge louse, and I recently won a battle for a sponge cavity that is home to a large harem of beautiful girls. But I'm starting to suspect that some of the girls are not what they seem: several look like men dressed as women. Am I being paranoid?*

*Hoodwinked in the Gulf of California*

Lots of men share your worries. Indeed, Julius Caesar divorced his wife Pompeia, because he heard she'd been seduced by a young man, Publius Clodius, who had dressed up as a woman to gain entrance to the Feast of the Good Goddess, a women-only event. Caesar was right about the disguise but wrong about the seduction. Yet he had no mercy on Pompeia, declaring, "Caesar's wife should be above suspicion."

Your fears are better-founded than his. Some male sponge lice look an awful lot like girl sponge lice. Take their size. Whereas you're probably twice the size of a female, these males are the same size as the girls. Or take their uropods—rear appendages special to sponge lice, shrimp, lobsters, and other crustaceans. Whereas you have enormous uropods that make you look as if you've taken a pair of curved horns and attached them to your back end, these males have small, dainty uropods, just like the girls. Thus incognito, these female look-alikes often infiltrate harems. The larger the harem the higher the risk, so if yours is large, I'm afraid it probably does hold some males disguised as females. And if you look carefully, I bet you'll find that's not all. If there are any "little ones," you'll find they are really men, too.

Why does this happen? Well, discretion is often the better part of valor, you know. In species like yours where females cluster together and thus one big male can easily defend a group, or in species where big males hold territories that females visit, small males cannot hope to compete directly. The big males would pulverize them. In such circumstances, therefore, it's common for some fellows to adopt more subtle tactics—a sneak approach.

There's a lot of this around. Sneaking is especially popular among fish, with reports of it in more than 120 species. Consider the bluegill sunfish, a freshwater species from North America. Big males defend territories that females visit to spawn. The male then looks after the eggs and the fry. Small males resemble females—and behave like females, flirting with the big males just as females would. When a real female comes along, the bogus female joins in the courtship and releases his sperm when the big male does. Meanwhile, even tinier males whizz out from the shadows at the critical moment. And Caesar thought he had problems.

But what determines whether a fellow will behave as a sneak? That depends. In some species, males adjust their behavior according to circumstance. Take the black-winged damselfly, *Calopteryx maculata*. Males defend territories along the banks of streams. But tired, old males can't do this successfully—young, energetic fellows chase them off. The oldsters don't despair, however. They'll sneak onto another male's territory, trying their luck with the girls while the territory owner is busy fighting or fornicating. Likewise, worn-out horseshoe crabs don't try to meet girls in the sea as younger males do. Instead, they rudely clamber on top of couples that are already spawning on the beach, adding their sperm to the mix.

Alternatively, during his early development a male might "decide" to become a sneak. For example, a male exposed to a poor environment during crucial periods of growth may switch off the genes that encode huge weapons or cumbersome ornaments. After all, weapons are expensive. There's no point in having them if you're never going to win a fight anyway.

This seems to be what happens in the bee *Perdita portalis*. The males either grow to be big and wingless, sporting huge mandibles, the better to fight with, or they remain small, grow wings, and don't bother with the mandibles. Which form they take depends on how much food their mother supplied them with. Males whose mothers provided lots of food become big. Males whose mothers didn't, don't.

Finally, sneaking might be genetically hard-wired—as it is in your case. The system is simple, based on one gene with three variants: alpha, beta, and gamma. Each sponge louse gets two copies of the gene, one from each parent. Alpha males arise when both copies received by a male offspring are the alpha variants of the gene. If a male has one copy of the beta variant, he will be a beta male—a female look-alike—regardless of what his other copy is. If a male has one gamma and one alpha variant or if he has two copies of the gamma variant, he will be a gamma male—one of the "little ones."

Female sponge lice don't care whom they mate with. Their priority seems to be female companionship. Thus, when females leave the algal fields where they've been grazing and swim to shallower waters to mate and brood their eggs, they are attracted to sponges that already contain other females. Because females tend to congregate, alpha males typically have either no mates or several. Which is why alpha males—in contests that can last more than twenty-four hours—fight to take over colonies that are already established. The entrance to a sponge cavity is a chimneylike structure. The resident alpha male lurks here, standing on his head so his huge uropods can stick out of the top. During a fight, the resident attempts to throw the intruder off the sponge; the intruder braces himself against the outside of the sponge and tries to haul the resident out of the chimney. Predictably, the intruder is more likely to be victorious if he's bigger than his opponent.

Beta males and gamma males are also drawn to sponges that already contain females. You shouldn't blame yourself if some gamma males have managed to infiltrate the sponge. I'm sure you've made a gallant effort to keep them out; whenever you've caught one, you've no doubt tossed him off the sponge. The problem is, you are standing on your head, attempting to catch these fellows by feel, and they are most persistent. After enough tries, they usually manage to slip past you and down the chimney. But the presence of beta males is galling. These fellows don't just look like females, they act like females, gaining access to the sponge by allowing you to court them. That's right—just like a female sponge louse, they allow you to shake them vigorously.

Why can't you at least spot the beta males and keep them out? Well, beta males are the least common type, comprising only 4 percent of the male population. (Gamma males are next, at 15 percent.) So perhaps it's not so bad admitting the occasional phony female. But, I hear you ask, won't this lead to the population's being taken over by transvestites? I doubt it. If one type of male had a clear and consistent advantage over the other two, it would rise in frequency in the population until the others vanished. Since we know all three types persist, each must do well in some situations but not in others.

A striking example of how a male's fortunes can depend on the other males around him comes from the side-blotched lizard, a small creature that lives on rocky outcrops on the coastal mountains of California. Here again, males come in three types: orange throats, blue throats, and yellow throats. Males with orange throats are big and belligerent and stake out large territories. Males with blue throats are smaller, less fractious, and hold smaller territories. Males with yellow throats resemble receptive females; these fellows neither fight nor hold territories. And as you'd expect, each type has a different style with the girls. Orange throats are wild philanderers. They copulate with all the females on their territory and often intrude into a neighbor's territory to copulate with females there, too. Blue throats are not interested in lots of mates. Instead, they are possessive fellows and guard their partners jealously. Yellow throats, as you'll have guessed, are sneaks. They have sex with females while other males aren't looking.

It turns out that each strategy is better than one other. The vigilance of blue throats means they are rarely cuckolded by yellow throats; however, blue throats cannot protect their females from the larger, more aggressive orange throats. Orange throats, meanwhile, are frequently cuckolded by yellow throats. The result? A perpetual oscillation. If orange throats do well one year, yellow throats do well the next, and blue throats the year after—and so on. For success, you've just got to be the right man at the right time.

༺

*Dear Dr. Tatiana,*

*I'm a peacock. But I have a lousy tail. It isn't very big, and some of the eyespots are wonky. The whole effect is walleyed. When I put my tail up, the hens don't even bother to feign indifference. They don't look at me at all. Is there anything I can do to impress them?*

*Invisible in Sri Lanka*

My advice: join a gang. If you can't make it on your own, gangs are often the solution. What sort of gang? Well, they vary, depending on the circumstances. In species where a few males hold territories and the other males loiter on the fringes, a typical gang takes the form of a loosely organized charge, the loitering males rushing a territory together. For example, in the bluehead wrasse, a coral-reef fish that lives in the Caribbean, young males sometimes team up to chase a large male from his territory. When a female arrives, all the young males spawn with her.

Similar invasions often happen in southern sea lion colonies. In this species, males are three times bigger than females and, as their name suggests, they also sport an impressive mane. The biggest males guard harems on the beach and chase the young males away. But young males are not without hope. From time to time, groups of as many as forty will rush the beach, breaking up harems and mating with females or even abducting them. To abduct a female, the male grabs her in his jaws, hurls her behind him, and then attempts to fend off her rescuers, often sitting on her to prevent her escaping.

In your case, though, such yobbish tactics won't work. Peahens don't sit about in harems. They are much more independent. Thus, the best gang for you would be a lek—a group of males displaying together.

Leks are common in species like yours, where females want nothing from males but their sperm. (Females can form leks as well, but they rarely do.) By definition, leks are not organized around food or nesting sites or anything else a male could usefully defend. Instead, a female visits a lek to compare and contrast, to see who's the hottest of them all. Having selected, she mates and goes away again. For a girl this a great system. She gets to have sex with the guy she likes best—and doesn't even have to see him in the morning.

But it's tough on boys. Being judged means you have to compete. That's why lekking species produce some of the most astounding shows of talent, the most bodacious beauty contests on earth. In these pageants, males are being rated not on their personalities or their skill with children but on looks, voice, agility, or whatever it is that females find sexy. In *Cyrtocara eucinostomus*, a fish that grows to ten centimeters (four inches) in length, what females find sexy is sand castles. The males build conical sand castles, one mouthful of sand at a time, at a rate of one mouthful every fifteen seconds. It's a massive task. The tallest sand castles—and females like the tallest ones best—take two weeks to build and are almost a meter wide at the base. That's nearly ten times the male's own length. In the hammerheaded bat, a large bat from West Africa, females prefer males who can honk loudly. This accounts for the male's peculiar appearance. Twice the weight of a female, he has a head that looks like a horse's and a voice box that occupies more than half his body cavity. During the breeding season, males assemble to honk for several hours every evening and again every morning, and the females take their pick.

But if being in a lek is so difficult, why am I suggesting you join one? Especially since you've already said you can't cut it. There are a couple of reasons.

The first is general. In many lekking species, females are attracted to groups more than they are to males all by them-selves—and the larger the lek, the more pronounced the attrac-tion. Since you obviously won't attract any females by yourself, you have nothing to lose by joining one. (Peacock leks, unlike leks in some other species, have room for fellows who lack sex appeal.)

The second is more specific to your situation. You can be valu-able if you find your brothers and half brothers and lek with them. Then, even if you personally never mate, you can help them attract mates just by being there. I know this sounds like a raw deal. But because you and your brothers share many of the same genes, you can pass on your genes by helping them to pass on theirs. So how will you find these brothers? Don't worry about that. The peacock brotherhood recognizes its own, even when brothers have never met before. I cannot tell you how. As in any good secret society, the answer is (for now at least) known only to peacocks.

In lekking with your brothers, you're way ahead of most other lekking species. The only other species where brothers are known to lek with brothers is another bird, the black grouse. And you're way ahead of other species' gangs of ill-assorted hood-lums. Ask any mafioso: whatever your purpose, gangs of brothers are the most dependable. Lions are the most famous example of fraternal cooperation. Lionesses live in family groups and rear their cubs in a joint litter. A lion's success in siring cubs depends on how long he can stay with a particular family of lionesses. And that depends on how many lions he's working with. The larger the coalition of lions, the longer they hold on to a pride. And it turns out that coalitions of more than three are always composed of littermates—brothers, half brothers, and an occasional cousin. (Pairs of lions may or may not be related; threesomes are either three littermates or two littermates and a stranger.) This means that brothers and cousins born at the same time—whether small or large, weak or strong—will have a much better chance of having children later on. Likewise, lionesses who give birth to several sons at once will have more grandchildren: their sons will be able to form larger coalitions.

Remarkably, lionesses appear to have evolved to do just that. Yes, they appear to be able to improve the odds that their sons will have coalition partners. Analyses of lion birth records show two significant patterns. The first is that when a female gives birth to a large litter, it will typically contain more sons than daughters. The second is that when several females in a pride give birth simultaneously, the cubs are disproportionately male. How do females synchronize the births? Remember that when a new group of lions takes over a pride, the first thing they do is kill or chase out any cubs that are around. This brings all the females back into heat. As a result, the females become pregnant at about the same time and give birth at about the same time, thus producing a large cohort. (Later in the males' tenure in the pride, births are less synchronous.) And sure enough, lionesses tend to have sons rather than daughters in the first pregnancy after a takeover. How they manage to do this is, however, unknown.

So you see, my walleyed friend, if your mother took a leaf from the lion's book, you won't raise your tail in vain.

ᔕᔕ

*Dear Dr. Tatiana,*

*Ma'am, I'm a field cricket, and I'm mighty annoyed. I've been singin' and singin', but there's not a gal in sight. The last one I saw went off with a creep who'd been skulkin' in the bushes, and I got left with nothin' but a few flies. Am I doin' something wrong?*

*Hoppin' Mad in Texas*

You've been a victim of the infamous "let the other guy pay for dinner, I'll go home with the girl" strategy. I'm sure everyone knows someone who gets away with this. I know I do. But the problem is most acute among species where males announce their whereabouts to any girls who might be passing by. While some males perform, others lurk. In some Caribbean ostracods—the shrimp that look like kidney beans—males give off sexy bursts of light as they bob in the water just after dusk falls. How do they make light? They secrete chemicals that react with the water to produce a glowing spot. As the male swims, he leaves behind a trail of light. Different species produce light at different rates—and leave different trails. If you were to run a net through the water column, however, you'd find a lot of males bobbing along without flashing, presumably hoping to intercept females attracted by the pulsations of signaling males.

Parasitism of this kind is obviously a tactic that, for its success, depends on the relative numbers of parasites and performers. You can't all be parasites: after all, someone has to sing for supper. In any event, given that there's a drawback to remaining silent—girls typically prefer singers to skulkers—why don't all males just sing?

For a number of reasons. Among bullfrogs, for example,

skulking is a matter of necessity, not choice. Bullfrogs breed in ponds, and the biggest males sing from the territories that they hold—and defend vigorously. If they see a trespasser, they wrestle him—and if they win, they teach him a lesson by holding him under water for a few minutes. (Why don't they drown him and be rid of him? Who knows. Perhaps they don't want rotting bodies fouling the water.) Since size is always an advantage in wrestling, the little guys rarely win. So they keep a low profile, hiding in the water at the edge of a territory with only their heads sticking out, ready to surprise females on their way to the big male. But among field crickets such as yourself, the reason for silence is different. It's those flies you mentioned.

Female field crickets much prefer singers. Unfortunately, so do females of another kind: parasitic flies. These deposit their larvae on the singer's body; the larvae burrow inside and eat away the singer's flesh. Death follows within a week. Whether a cricket becomes a singer or a skulker has a strong genetic component— some males are naturally less inclined to sing than others. But whether skulkers fare better than singers depends on the abundance of the parasitic flies. In years when flies are abundant, skulkers do better and their genes spread. In years when flies are scarce, singers do better and their genes spread.

It's a nasty, brutish world. Fringe-lipped bats pick off calling frogs. The bats know their stuff—they can tell the difference between species that are poisonous and those that taste nice, and between those that are too big to handle and those that are just right. Little blue herons go after calling male crickets. Mediterranean house geckos sit silently outside the burrows of male decorated crickets—and snaffle up any females who come to hear the song. And so it goes.

Even worse, some predators don't just exploit these mating signals—they mimic them. Bolas spiders, for example, are chemical chameleons. Adult females, which are sedentary and plump and have coloring reminiscent of bird droppings, have evolved an unusual hunting technique. They catch prey by swinging a kind of lasso. The lasso is made of a sticky ball attached to the end of a thread, which the spiders use one of their front legs to swing at passing insects. In some species, the spiders even whirl the lasso around their heads as if they'd stepped out of a spaghetti western. How do they bring their prey within range? When they are hunting, they give off the scent of female moths. Sure enough, male moths come flying over to investigate. Poor fellows. If they are caught, they are wrapped in silk and hung from a thread until the spider is ready to eat them. (Different species of spider give off different scents—and attract different species of moth.)

Remarkably, both juvenile and male bolas spiders also hunt by giving off a female scent—although this time, by an amusing coincidence, the scent is that of female moth flies, flies that look like miniature moths. Why do males hunt like juveniles? Males typically grow little after hatching (and are thus far smaller than their mates). Males and juveniles do not use a lasso, probably because they are too small to manage it effectively. Instead, they capture prey using their four front legs.

Occasionally, though, the prey bites back. Take the digging wasp *Oxybelus exclamans*. These wasps are liable to have their nests taken over by an insect version of the cuckoo: a flesh fly that lays its eggs in the cells of a wasp nest. When the fly larvae hatch, they eat all the food that the wasp had provided for her own children. Male flesh flies loiter around wasps' nests in the hopes of meeting female flies. If they're not careful, though, these fellows will find that the wasp has captured *them* for her children to eat.

As for *your* problem, if you see another of those flies, I urge you to go straight to the doctor. You may need major surgery.

⤳

*Dear Dr. Tatiana,*

*I'm a marine iguana, and I'm appalled by the behavior of the young iguanas of today: I keep encountering groups of youths masturbating at me. It's revolting. I'm sure they didn't dare act this way in Darwin's time. How can I make them stop?*

*Disgusted in the Galápagos*

I get a lot of letters from young male marine iguanas, frustrated because the girls ignore them. But this is the first time I've heard complaints from the other side. Look at it from the guy's point of view. Here he is, a tasteful shade of red, his spiky crest a full twenty centimeters from his crown to his tail—he's ready to go, desperate to use one or the other of his penises (yes, like many reptiles, he has two, a left and a right penis). But being young and therefore small, he doesn't have much of a chance. It isn't just that you ladies prefer to mate with older, bigger males. It's that even if he manages to mount a female, the odds are he'll be shoved aside by a bigger fellow before he climaxes. That's why young males masturbate when they see a girl go by. Wanking reduces the time they need to ejaculate during sex—and thus reduces the risk of being interrupted before their climax. So I'm afraid the behavior may be here to stay. Young wankers probably sire more children than those who abstain.

Does anyone else masturbate? Yes. In many primates, individuals of both sexes masturbate a lot. Take the sooty mangabey, a smoke-colored monkey from West Africa with a long tail and extravagant tufts of whiskers on its cheeks. Some females use their hands to stimulate themselves during sex. Male and female orangutans stimulate themselves with sex toys they've made out of leaves or twigs. One female chimpanzee who was raised in a human household masturbated to a copy of *Playgirl*, thrilling to the photos of naked human males, especially the centerfold. Other mammals masturbate too. Male red deer do it by rubbing the tips of their antlers through the grass. The whole act takes fifteen seconds from start to spurt, and during the breeding season some stags masturbate several times a day. But does anyone else do it, like marine iguanas, out of fear of being interrupted in bed? Frankly, the matter hasn't been the subject of much research. There has been more work on a related topic: big balls.

Big balls are a more conventional way for small males to increase their odds of fertilizing eggs. The logic is simple. In species where small males have to sneak to mate, they are guaranteed to be at risk of sperm competition. As you know, sperm competition is often like a raffle—more tickets, more chances. Therefore, small males who invest a larger proportion of their bodies in making sperm can buy more raffle tickets—and better their chances of success whenever they mate. Meanwhile, large males, as long as they are reasonably effective at guarding females, don't need so many tickets or such big tackle.

That's why there's often no relation between the dimensions of a man and the dimensions of his privates: bigger men do not necessarily have bigger bits. Indeed—more's the pity—it's often the opposite. The plainfin midshipman, otherwise known as the California singing fish, takes this to extremes. Males have either big brains or big balls. The brainy kind excavate cavelike nests beneath rocks in the intertidal zone. Once a male has prepared a nest, he hums to attract females. A single humming bout can last for a quarter of an hour. Thus, he has large muscles for humming and extra neurons to control the muscles. When a female arrives, she slowly lays her eggs on the ceiling of the nest; as she does this, the male quivers beside her every few minutes, a sign he is releasing sperm. When she's done—which can be as long as twenty hours after the laying of the first egg—he throws her out of the nest so that he can guard the fertilized eggs and sing to attract more females.

The other type of male—the one with the big balls—sneaks into the nest at the crucial moment. These fellows can't hum: they lack the mental and physical apparatus. The best they can do is grunt. But boy, are they well hung! As a proportion of his weight, a sneak has gonads nine times heavier than a brainy male does. His gonads are so large that his stomach bulges as if he's pregnant. No wonder he grunts.

So you brainy types out there shouldn't feel too smug. Your position is only safe if sneaks are rare. If sneaks are common, then you're at greater risk of sperm competition and should invest more in making sperm. Thus you should have larger balls. Compare two species of dung beetle, *Onthophagus binodis* and *Onthophagus taurus*. These are among a score of dung beetle species introduced to Australia from other parts of the world. You see, Australia had no native cows, and so when humans imported them, the cows produced more dung than the native dung beetles knew what to do with. As a result, large quantities of cow manure accumulated in pastures. To solve this problem, dung beetles with a talent for disposing of cow dung were invited to immigrate. Which is to say, dung beetles were captured in other parts of the world, quarantined—I love the thought of an insect quarantine—and then released in Australia.

To return to the matter at hand, *Onthophagus binodis* and *Onthophagus taurus* have similar—and by now, familiar—biology. Males again come in two sizes, big ones with horns, little ones without them. (In *O. binodis*, the big males have a single horn on their backs; in *O. taurus*, as the name suggests, they have two curved horns on their heads.) Males and females meet at fresh mounds of dung. Females pair up with big males, and together the pairs dig burrows, pausing from time to time to copulate. (Copulation in dung beetles has rarely been observed, but in *O. binodis*, the male caresses the female with his first two pairs of legs, mounts her, and then goes into spasms during which he taps her back with his front legs.) A typical burrow has several passageways branching off from a central corridor. At the end of each passageway, the beetles deposit a wad of dung. The female lays a fertilized egg on the wad and then seals off the passageway with earth. Although the male will give his partner lots of help collecting dung and so on, he does not like to leave the entrance of the burrow unattended. This is a wise precaution, since the small males rush into the burrow when the big one isn't looking and copulate with the female. Sometimes, the small males are even sneakier and dig their way into burrows, erupting through one of the walls.

Sneak attacks are more likely, however, to be a problem for *O. taurus* than for *O. binodis*. In *O. binodis*, sneaks constitute only a third of the male population. Any particular big male is therefore at low risk of cuckoldry. And sure enough, in this species, a little male expends more on sperm production than a big male does, having both larger testes and a higher sperm count. By contrast, in *O. taurus*, almost two-thirds of males are sneaks. Accordingly, in this species, big males and little males are indistinguishable on the basis of balls alone.

∽

A parting word to all you guys who worry about being small or looking unimpressive (and girls, too, take note). In many, many species, males fall into distinct types. Physical characteristics are coupled with personality traits, which means that you can tell how a fellow will behave just by looking at him. The number of types and their particular attributes vary from species to species.

Two types, however, are especially common: the Hunk and the Runt.

The Hunk suffers from a God complex: he has a high opinion of himself, he's always keen to fight, and he spends lots of time strutting and preening. He often has many girlfriends, but he would be horrified at the thought that one of them might cheat on him. And although he's handsome, he has, alas, small private parts.

The Runt is self-effacing in groups of other males. He dislikes fighting, but he's pushy with the girls. He is not to be trusted— he never commits to one woman and he's not ashamed to cheat on his best friend. But here's the thing: physically puny, he often has big parts. Runts make love, not war.

# PART II

# THE EVOLUTION OF DEPRAVITY

Every war has atrocities: the war of the sexes is no exception. The more her desires clash with his, the more diabolical the outcome.

## ·6·
## HOW TO MAKE LOVE
## TO A CANNIBAL

---

**R**ule number one: Never get eaten during foreplay.

---

*Dear Dr. Tatiana,*

*I'm a European praying mantis, and I've noticed I enjoy sex more if I bite my lovers' heads off first. It's because when I decapitate them they go into the most thrilling spasms. Somehow they seem less inhibited, more urgent—it's fabulous. Do you find this too?*

*I Like 'Em Headless in Lisbon*

Some of my best friends are man-eaters, but between you and me, cannibalism isn't my bag. I can see why you like it, though. Males of your species are boring lovers. Beheading them works wonders: whereas a headless chicken rushes wildly about, a headless mantis thrashes in a sexual frenzy. Why can't he be that way when he's whole? Well, it's hard to have wild sex if you're trying to keep your head.

A male praying mantis is in danger during his approach and his departure, but while he's actually on your back—the position in which intact males have sex—you cannot attack him. However, you do not need him intact to have sex with him. If you rip his head off on the approach, his body will go into spasms that allow his genitalia to connect with yours. Unsurprisingly, though, he does not want to have his head removed. Put yourself in his place—you'd be trembling to the tips of your antennae. If you so much as glimpsed a female, you'd freeze. And then you'd start to play grandmother's footsteps. Whenever she looked away, you'd creep closer. Whenever she turned toward you, you'd stand like a statue—no, no, don't look at me, I'm just a leaf—for hours if you had to. The aim? To get close enough to leap on her back. Once on her, you could make love unmolested. But a single false step, and you're at the pearly gates with your head tucked under your arm. In grandmother's footsteps, the stakes don't get higher than that.

Females in more than eighty other species have been caught eating their lovers before, during, or after sex. Spiders are the most common culprits, although several other mantises, some scorpions, and certain midges also number among the guilty. The midges—tiny flies with big appetites—dispatch their lovers in a particularly horrible way. The female captures her mate as she would any old prey and plunges her proboscis into his head while they link genitalia. Her spittle turns his innards to soup, which she slurps up, drinking until she's sucked him dry, then dropping his empty shell as casually as a child discards a dull toy. Only his manhood, which breaks off inside her, betrays the fact that this was no ordinary meal.

But perhaps there's an innocent explanation for this behavior. Perhaps it is a regrettable but genuine mistake. Or perhaps it is a rare psychosis brought on by life in captivity. After all, roughly a third of cannibal species have been seen at it only in the laboratory: perhaps it happens because in a small cage the male can't run and can't hide. Perhaps. But the European praying mantis is one of the few species that has been watched in both the laboratory and the wild—and cannibalism is equally common in both places. The difference is that laboratory sex takes several hours longer, apparently because the male is too terrified to dismount. (Normally, when the male is done, he drops into the undergrowth, putting himself out of reach. Laboratories usually don't feature undergrowth, and the male stays where he is, as if pondering his predicament.) As for "I ate my lover by mistake"— well, I can't say accidents never happen. But I know of several spiders where there can be no doubting the female's intention to take head, not give it. When she sees a male, she beckons him over and adopts a submissive, "I'm yours" posture—only to pounce on him, wrap him up, and store him in her larder before he can say "cannibal."

The trouble is, all too often the male is captured and eaten before he's had a chance to mate. From his point of view, this is a disaster. If he's eaten during foreplay, that's it: his genes get naturally deselected from the population. And from her point of view? The habit is not as self-defeating as you might think. For many of these creatures, a male represents a substantial meal. A female garden spider, for example, becomes noticeably plumper with each male she consumes. Her only risk is that she'll be so aggressive toward her suitors that she'll die as she lived—a grumpy old maid. But that risk is negligible.

To see why, let's step back and take a more general look at what happens when females routinely attempt to eat their lovers before sex. First, imagine a land where all the girls are equally rapacious. And imagine that each girl will meet just one boy in her whole life. If every girl eats her only suitor instead of screwing him, everyone loses: nobody reproduces and the population goes extinct. However, what if some boys could somehow escape being eaten, at least until they had done the job? Any boy able to escape would have a huge advantage over those who were not. And if the trick to escaping had a genetic basis, genes for escaping would spread. After all, every male in the next generation would be the son of a successful escapee—and thus females would be mated in spite of their rapacity.

Granted, in real life some girls may not be so fierce. This complicates matters. Girls who don't eat their mates are at no risk of remaining virgins, so if everyone has only one suitor, kinder, gentler females will win the day. This is because, with noncannibals in the population, a male lucky enough to meet one will be able to mate even if he doesn't have escaping genes. As a result, the advantage to being an escaper will be smaller, genes for escaping will spread more slowly—and rapacious females will be more likely to encounter males that cannot elude them. Having eaten their only mate, rapacious females will leave no offspring, and genes for rapacity will disappear.

Now add another dollop of reality to the situation and consider what happens if each female is likely to meet lots of males. In this scenario, it won't matter to her if she eats most of them. Indeed, girls, it may count against you if you don't try to. That's because if everyone else is attempting to catch and eat their lovers, attempted cannibalism becomes a test. In a man-eating culture, your sons will survive and reproduce only if they can evade a female's clutches, so it would be wise to check their dads' abilities. At the same time, any male who can escape will again have a huge advantage over those who cannot, and genes for escaping will spread.

To sum up, the more likely females are to try and eat their mates, the larger the advantage in being an escapee and the faster the population will be made up of great escapers. In most situations, then, you should expect to see not cannibalism but escape.

But how does a male get close enough to copulate while avoiding capture? Grandmother's footsteps is one technique—but that won't do if he has to tiptoe across a spiderweb, where every twang on the threads tells the owner where he is. Besides, while a male mantis can always crouch at the end of his partner's back, spider sex is more perilous. A male spider has two penises (called pedipalps), one on each side of his mouth. A female spider has two genital openings on the underside of her belly—you see the difficulty. It is impossible to have sex without getting up close and personal.

The most reliable way to escape unscathed is to disable her somehow. That's why male *Tetragnatha extensa* spiders are not afraid of sex: they have spurs on their fangs to wedge open the female's jaws so she can't bite during their embrace. The male crab spider *Xysticus cristatus* is a great lover: he goes for bondage, tying the female down (lucky her!) before making love. And in *Argyrodes zonatus*, a tiny silver spider that dwells on the webs of much bigger spiders, the males are nature's frat boys. On their heads they have a horn that secretes a powerful drug. They offer the horn to the female to suck so she'll get high and won't be able to resist their advances. Better hope she doesn't wake up with the munchies . . .

As for Mr. Praying Mantis, he's had a stroke of bad luck. When he's possessed of his head, his brain sends messages to his private parts telling him how to behave. This holds his libido in check until he's in position. When he loses his head, the messages that inhibit sexual behavior cease—and he turns into a sex fiend. The result is that he can copulate when there's almost nothing of him left. Yet although this sounds like proof that he has evolved a spectacular adaptation to being eaten, the "lose head, have sex" reflex is actually rather common among male insects. Something analogous even happens in humans: throttle a man and like as not he'll get an erection, not from erotic pleasure in dying but because "Down, boy" signals from the brain stop coming. For most fellows, such a reflex is simply a medical curiosity. But most fellows don't have to face Ms. Mantis in the bedroom.

ᐁ

*Dear Dr. Tatiana,*

*I'm an Australian redback spider, and I'm a failure. I said to my darling, "Take, eat, this is my body," and I vaulted into her jaws. But she spat me out and told me to get lost. Why did she spurn the ultimate sacrifice?*

*Wretched in the Wilderness*

What could be more perverse? A known man-eater refusing to eat a man who wishes to be eaten? Needless to say, your problem is unusual. But so are you. When faced with making love to a cannibal, most males do not try to make themselves more delicious.

In the first place, although being eaten after sex beats being eaten beforehand, most guys prefer not to be eaten at all. No surprises there: death puts a stop to amorous adventures. Anyone who has a good chance of mating again should hump and run— no whispering of sweet nothings, no postcoital cigarette. If anything, you should pummel her a little to stop her chasing you. In the scorpion *Paruroctonus mesaensis*, the male whacks his partner several times before racing off; in the wolf spider *Lycosa rabida*, the male tosses his lover in the air, leaving her in a crumpled heap as he hurries away.

But what if you haven't a hope of mating again? This would be the case if, for example, you were only capable of mating once, or if your life expectancy were short, or if a quest for another female were sure to fail. Then, as long as you've accomplished your mission, you shouldn't fuss if your lover eats you. The male spider *Argiope aemula* vigorously resists capture before he's had sex— but the excitement is usually too much for him, and he expires *in copula*. It's fine with him if she opts for gastronomic burial. In the bristle worm *Nereis caudata*, something similar goes on but for once it's the man who eats his wife. These worms, which look like bottlebrushes, live in sand and mud on the seafloor. Once the female has laid her eggs, the male puts them into a long tube that he makes, and fertilizes them. He tends the eggs, like a dragon guarding treasure, until they have hatched and the larvae are ready to go out into the world. Shortly after the female lays her eggs, she gives up the ghost. If her mate decides, as he sometimes does, to hasten her end by having her for lunch, so be it. It's all the same to her.

Do other males eat their mates? I have never heard of it. But note: this is not to say males don't eat females. They do. Just not during sex. Platonic cannibalism is a problem for creatures from apes to amoebae. It's depraved out there. The sand shark, for example, practices intrauterine cannibalism. That's right, the biggest fetus gobbles up its embryonic brothers and sisters while they are in the womb. Surely you know the rhyme:

The shark, he is a vicious beast,
Tears fin from fin at every feast.
But it's no surprise he should do so—
He ate his sibs *in utero*.

The reason platonic cannibalism is so much more common than the sexual kind is simple. Cannibalism is risky: your intended victim may, at any moment, capture you. Most cannibals, therefore, are cowards and never pick on someone their own size. In a typical cannibalistic society, adults eat children, big children eat small children, and small children eat eggs. Even among amoebae, cannibals are giants. So you see, cannibalism between adults—of any kind, sexual or not—is rare.

Moreover, for most males, it makes no sense to eat their mates: they'd lose the eggs they've struggled to fertilize. That's why the male paddle crab *Ovalipes catharus* is a gentleman. In this species, everybody eats everybody, with one caveat: you are at risk only while molting. That's because the Coward's Rule holds: molting crabs can't defend themselves. For several days they have neither a shell that can withstand blows nor claws that can deliver them.

Unfortunately, however, as is common for crabs, female paddle crabs must mate while molting. This leaves them vulnerable to cannibalistic attack. But help is on the way. On meeting a female who's about to molt, a male picks her up and carries her until she's gone soft. He then makes love to her ever so slowly—sometimes taking several days over it—and protects her from males with less honorable intentions until she's hardened again and can look after herself. This gallantry is hardly selfless, however. By fending off crabs who might eat her, he also fends off males who might ravage her and thus raises the odds he'll be the only dad for her current batch of children. It's quite a prize: large females carry more than 250,000 eggs per batch.

If males don't usually eat their lovers, what about hermaphrodites? Almost nothing is known about this matter. Many hermaphrodites are platonic cannibals, though, and I can imagine that sometimes, when two individuals meet, one wants nooky but the other wants lunch. Since both parties would want to avoid being eaten, I'd be surprised, though, if it's a common problem. My guess is that whenever sexual cannibalism could be a risk, countermeasures rapidly evolve. For example, if you are most vulnerable when mating with someone bigger than yourself, you might evolve a horror of larger individuals. It has even been suggested that the risk of cannibalism explains the lightning-fast sex of the hermaphroditic sea slug *Hermissenda crassicornis.* "Slug" hardly conveys the ethereal beauty of this animal. No more than a couple of centimeters long, it looks like a land slug fallen under a powerful enchantment: a delicate, glassy body tinged with pale blue, a fetching orange stripe running down the back of its head, and, lower down, a glorious mass of feathery protrusions as if the animal were wearing a coat covered with foxtails. But when it comes to sex, these creatures don't stand on ceremony. Like knights jousting, the two animals whizz past each other without stopping, lances held out in an effort to knock the other up. Yet although this custom is mighty peculiar, it's not clear that cannibalism is the true cause.

Speaking of peculiar, nothing is as peculiar as you male redback spiders wanting to be cannibalized. The urge is so strong that you sometimes fight for the privilege, one male snatching a rival out of the female's jaws, bundling him in silk as if he were a fly, and then marching into the jaws of doom in his stead. The scene is all the more absurd because you fellows are midgets—one hundred times smaller than the female—so a fight looks like two rabbits dancing around a lion. It goes without saying that such a death wish can evolve only in special circumstances. That is, being eaten must mean you leave more offspring than if you are spared. So far, your species is the only one known to meet this criterion. A male redback who gets himself munched fertilizes more eggs than a male who survives. Why? Remember that spider sex means inserting your pedipalps into the twin orifices on her hairy black underbelly. But even with the tip of your abdomen in her jaws, you can still reach the orifices. And it turns out that sex takes longer when she's chewing away on you, which gives you the chance to deliver more sperm and thus fertilize more eggs. So your challenge is to make yourself more appetizing.

The secret is picking your moment. Female redbacks aren't greedy; when they're not hungry, they don't eat. If you offer yourself right after she's feasted, forget it. You've got to wait until she gets that mean and hungry look in all eight of her beady little eyes. And then, for what you are about to receive, may your kiddies be truly thankful.

༜

I'm afraid we don't know why females in some species become man-eaters while their sisters in closely related species do not. All we can say—and it seems obvious—is that, without exception, sex cannibals are never vegetarians but always predators and that they tend to be larger and stronger than their victims and therefore able to overpower them. Boys, if you have fallen in love with a large, predaceous lady who tries to bite your head off, you may be the dinner as well as the date. If you suspect you could be at risk, you must ask yourself this: do you want to meet your maker now or later?

If the answer's later, then think SAFE SEX: Stealthy Approach, Forceful Embrace, Swift EXit.

If the answer's now, think again: are you mortally sure you will be rewarded? If so, then prepare your last words—and pray your epitaph will be "He was fruitful."

Girls, eating men without screwing them is just plain wrong. But hey, you only live once. If you like making mincemeat of your lovers, remember that cannibalism is the right choice if and only if you run little risk of remaining a virgin. If that's taken care of—bon appétit!

# · 7 ·
# CRIMES OF PASSION

---

**M**urder, wife beating, rape. Why do they happen? Because some boys won't take no for an answer.

---

*Dear Dr. Tatiana,*

*I'm supposed to be a solitary bee, but I can't get any time to myself. Whenever I poke my proboscis out of my nest, I'm hounded by guys who apparently have nothing better to do than make a nuisance of themselves. They think it's funny to molest me as I'm doing my chores. It's not funny. It's maddening. How can I get them to buzz off?*

*A Girl's Never Alone in Oxford*

Male bees and wasps are notorious layabouts. Look at social species—species, such as honeybees or hornets, that live in large nests ruled by a queen. Males laze around while their sisters—

the workers—do the drudgery of gathering food, cleaning the nest, and rearing the young. (Occasionally, workers rebel. In the paper wasp *Polistes dominulus*, workers indulge in "male stuffing." It's a cruel sport. When a male gets stuffed, he's bitten, kicked, and shoved headfirst into an empty cell in the nest. The stuffer keeps pushing and biting his rear end to stop him from coming out. A male is most likely to get stuffed when a worker arrives back at the nest with food; by the time he's managed to unstuff himself, the food will have been given to deserving members of the nest, like grubs.)

But although social species, with their humming hives, are the darlings of the media, most bees and wasps hail from solitary species like yours. In solitary species, each female reproduces; there are no workers. But there are still layabouts. All too often the female is a supermom—building the nest, laying eggs, supplying provisions for each grub to eat when it hatches—while the male is a feckless deadbeat. This can cause problems. In your case, it means that after a male has breakfasted on nectar, he has a whole day ahead of him with nothing to do but chase the girls.

Yes, it's a bore. Males of your species are not subtle: they start their seduction with a pounce. If a male pounces on you while you are flying, he can knock you to the ground; if you have to fly a gauntlet of males, you could be knocked to the ground once every three seconds, making it a real challenge to collect pollen and nectar to provision your nest. Nevertheless, you should thank your lucky stars. In other species, when idle youths hang about in packs things can get much, much worse.

A female mountain sheep may be chased for miles by packs of eager young rams. This is exhausting and potentially dangerous: to escape she'll often jump to narrow ledges on the cliff face. Still, I bet she's glad she's a mountain sheep. The Île Longue, the biggest island of the Kerguelen archipelago, a cluster of glorified rocks just north of the Antarctic Circle, is home to a flock of domestic sheep that have been left to their own devices for over thirty years. The result? It makes *Lord of the Flies* look like a teddy bear convention. Ewes are not merely chased but battered to death by gangs of rams. A victim will be pursued until she's too tired to flee, then the rams will try to mount her. They rarely succeed—as one climbs on, another charges and knocks him off—but that's no comfort. Sessions can last for hours, the ewe getting more and more bashed; if she falls in exhaustion, she'll be kicked and butted until she gets up. If, at the end of it all, she's not dead, she risks being finished off by giant petrels. These huge seabirds—they boast a wingspan of more than two meters—have the disgraceful habit of disemboweling weak animals, punching them in the anus with their heavy, hooked beaks.

Ewes are not the only ones to draw the short straw. Female frogs can have a terrible time. Take the quacking frog, an Australian species named for the quacking sounds that males make to attract mates. Like many other frogs, females come to pools to spawn. If all goes well, girl frog meets boy frog, he climbs on her back and wraps his arms around her—a classic froggie hug called amplexus. She releases eggs into the water; he squirts them with sperm. But if she has the misfortune to attract several males, they will push and shove as they jockey for position. At best, fewer of her eggs will be fertilized; at worst, she'll asphyxiate in their embrace. Such tragedies are not unique: female wood frogs who attract several males occasionally drown in the ensuing melee.

But better drowned than dismembered, I say. In some solitary bees and wasps, males emerge from the winter earlier than females do and gather around burrows that females will emerge from. When a female does emerge, she may be dismembered and killed as the boys struggle to possess her. Female yellow dung flies go to fresh cowpats to mate and lay eggs. If they get set upon by several males they, too, may be torn apart—or drowned in runny excrement. Perhaps dismemberment is better than I thought.

Meanwhile, once a year, female northern elephant seals gather on beaches to give birth, nurse their pups, and fornicate before going back to sea. On many beaches, the biggest males—and when I say big, I mean five meters (sixteen feet) long and weighing over two and a half tons—can keep smaller males out of the main female arena. But these young males are raring to go. If they spot a female making her way back to the sea, they will gallop over to her—and in their desperation to mate, may batter her to death and then fight for the corpse. Hawaiian monk seals, the only wholly tropical seals, are one of the most endangered seals in the world, with perhaps six hundred animals left in the wild. Unfortunately, the biggest threat to their survival is . . . other Hawaiian monk seals. The main cause of death is attacks by adult males on adult females, who are often mobbed for hours by gangs of amorous males, during which they may be battered to death or badly bitten and then eaten by tiger sharks drawn to the commotion. The problem is so acute that male seals are being rounded up and given drugs to suppress their libidos in an attempt to save the species from extinction.

All this must sound crazy. But the point is that none of the violence is intentional. The boys don't mean to hurt the girls: it doesn't do them any good if a girl dies before she's had their children. So why are they so aggressive? It's a catch-22. Things are most likely to turn sour when crowds of males hang about, all vying for favors from a few passing females. This is particularly likely if males congregate at places females must come to in order to breed or if a few males hold large breeding territories and the others lurk on the fringes. Then, it's nothing ventured, nothing gained: if a fellow stands aside and doffs his cap to a girl, he's guaranteed not to mate with her. Instead, he should rush in and press his suit. If he can beat off other contenders, he'll have the chance to fertilize her eggs. The trouble is, the same is true for everyone—hence the scrum. If she dies as a result—well, that's too bad. From his point of view, it's as though someone else won her.

From her point of view, the situation obviously looks different. No one wants to be dismembered, drowned, or battered to death, not least because death puts the kibosh on future reproduction. If the risk of coming to an untimely end—or being badly hurt—is appreciable, you'd expect females would evolve countermeasures. It's the usual story. Some females may have attributes that make them less likely to be hassled, hurt, or killed. If these attributes have a genetic basis, the genes involved will spread. Having said that, when a girl is facing many males, there may be little she can do by way of self-defense. Even the greatest warriors can't usually fight off several attackers at once or fight for days without rest.

So what works? A popular solution is to hire a bodyguard. Female yellow dung flies, for example, prefer to mate with large males and are more likely to be left in peace if they do so. Female northern elephant seals who copulate with someone on their way to the ocean are then escorted down the beach—and are less likely to be attacked by anybody else. Among water striders—seemingly delicate insects that skate about on the surfaces of ponds and streams—a female will fight to rid herself of a pesky male if few males are around. If, however, fighting off one fellow will just mean hassle from another and another and another, she accepts whoever comes along first. His presence will squelch the ardor of others. Bodyguards can be useful even when one's life is not at risk: when escorted by a male, female pheasants and pigeons spend more time eating and less time checking for predators or fending off other males than when they are on their own.

But you, my busy friend, don't need a bodyguard to protect you from harassment. In your species, males buzz about at predictable times and places. Better yet, there is no single place that you must go to. This means boys can't lie in wait at a place you can't avoid, so you can simply steer clear of them. If they're out in force at one bed of flowers, visit another. That's what most girls in your species do.

And cheer up: there are occasions when the pack mentality can be turned to a girl's advantage. Take the beewolf *Philanthus basilaris*. Beewolves are solitary wasps that hunt other bees and wasps. The males cluster together, each defending a small territory and making a pest of himself by chasing every passing insect in the hopes of finding the girl of his dreams. Poor bastards. They think they're going to get laid, but instead they get laid to rest. Females exact a grisly revenge for unwanted attentions. Girls who have already mated sometimes visit male aggregations not to make dreams come true but to provide for their families: they come to collect males to put into their nests for the grubs to eat when they hatch. With males so eager to mate, it's easy pickings. And a fellow who falls into the arms of one of these dames will discover a fate worse than death. He'll be stung until he's paralyzed but not dead—grubs like fresh meat—and then sealed into the nest until the grubs are ready to eat him alive. Revenge, it seems, is a dish best served cold.

✎

*Dear Dr. Tatiana,*

*I'm an Australian seaweed fly, and I'm a Sensitive New Age Guy. I know that no means no—but that doesn't get me anywhere. The girls in my species are tough Sheilas: whenever I make friendly overtures I get beaten up. Why are they so hostile, and is there anything I can do about it?*

Mr. Nice Is Mr. Frustrated in Mallacoota Bay

To hell with political correctness. In your species, no means yes. The girls are aggressive because they want you to overpower them; indeed, they refuse to have sex with anyone who cannot. So if you want to get some rumpy-pumpy, you're going to have to put up with being kicked and bucked until your partner submits to you.

For some creatures, violent sex is the norm. Crabeater seals—which, incidentally, eat not crabs but krill, small shrimp that live by the hundreds of millions in the chilly waters around Antarctica—bite each other savagely during sex, and both animals often end up drenched in blood. But although this looks nasty, it's like a human raking fingernails down her lover's back: the wounds aren't serious and there doesn't seem to be anything sinister going on. Why do they do it? Well, there's no accounting for taste.

A taste for violence has its hazards, of course. Rough love leaves many animals scarred by the experience. Elderly bull crabeater seals have scars all over their heads from years of female affection. When a bison bull mounts a cow, his front hooves strike her back, sometimes removing bits of skin. By the time she celebrates her eighth birthday, she'll probably have a couple of bald spots as a result. (During copulation, she has to bear the full weight of her massive lover: as he ejaculates, he goes into a spasm that brings his rear hooves off the ground.) Among dugongs, vegetarian sea mammals that snuffle around beds of sea grass using horny pads on their lower lips to yank plants up by the roots, females occasionally sport scars on their backs from the male's clumsy lovemaking.

Worse, accidents happen. In the pygmy salamander, the male plunges barbed teeth into the female's neck before presenting his packet of sperm; on occasion, he gets stuck and cannot tear himself away. Among southern elephant seals—the males are even bigger than northern ones, reaching six meters (twenty feet) and weighing about four tons—a careless bull may bite the cow on the head instead of the neck, killing her by crushing her skull. Dog mink occasionally make a similar mistake, piercing the base of the bitch's brain instead of grabbing the scruff of her neck. A male sea otter's idea of a kiss is a bite on the nose; this can be lethal if the wound gets infected.

Such disasters grab attention: bad news always does. But it's important to keep things in perspective. The crucial difference between these accidents and those at the hands of a libidinous mob is that these accidents are not a result of males acting in their self-interest. A male who goes about crushing his lovers' skulls won't leave many offspring. Thus, in species where girls are unlikely to be assaulted by a gang, the risk of being killed during sex will generally be trivial in comparison with the risk of dying in other ways, and avoiding such accidents will be a matter of luck, not evolution. That's because genes that could protect you from having your skull crushed—for instance, genes that build stronger skulls—won't spread if skull crushing is vanishingly rare.

But although dead girls can't have kids, injured girls might—it depends on the injury. So perhaps these accidents are more pernicious than they seem: what's to stop a fellow from thrashing a girl within an inch of her life, especially if a walloping puts her off sex and stops her cheating on him? Girls, fear not, they wouldn't get away with it for long. If, for whatever reason, lovemaking carried a real risk of serious injury, you would again expect the rapid evolution of traits that protect you from harm.

Consider two examples. The first comes from sharks. In these monsters, sex is often brutal. The male grabs the female in his jaws while attempting to insert one of his penises, or "claspers," which are pelvic fins rolled into tubes. (Males have two claspers. They lie parallel to his belly, one end touching his genital opening, the other poised to insert into the female. During copulation, sperm from the genital opening enter a clasper and are propelled into the female with a jet of seawater.) Grown-up females are often scarred or missing bits of fin. However, although I wouldn't want to be bitten by a shark—even one in a tender mood—if you're a girl shark, it's not that bad. Take the blue shark. Mature females often have fresh bites or scars on their bodies. But they can handle it: when females go through puberty, their skin starts to thicken up; by the time they are adults, they have skin twice as thick as that of a male of the same size—and more telling, their skin is thicker than his teeth are long. In the round stingray, a cousin of sharks, not only does the female have thicker skin, the male has special pointy teeth to hold her with. Whereas females and young males have an array of smooth teeth that fit together like the stones on a well-paved street, a mature male has spiky teeth to give his love bites more oomph. And sometimes it's the female who bites the male—and the male who has evolved to reduce the risk. *Falcatus falcatus* is a shark known only from fossils that are about 320 million years old. The male would be laughed out of the water by today's sharks: on his head he was adorned with a large handle, a modified fin that curved forward over his head, giving him the appearance of a shark-shaped flatiron. Judging by the compromising position in which one fossil couple was found, it looks as if the handle evolved for the female to bite during sex.

My second example of the evolution of countermeasures comes from the marine flatworm *Pseudoceros bifurcus*, a hermaphrodite.

Among hermaphrodites, every member of the species is a potential mate; conventional fighting between rivals competing for mates does not exist. Indeed, for many hermaphrodites, sex is an amorous, drawn-out affair. But not for *Pseudoceros bifurcus*. In this species, individuals apparently prefer the male role over the female role, for they have evolved a technique of hit-and-run insemination. This involves stabbing the penis anywhere into the victim's body before gliding off with all possible speed. But since being stabbed in this way inflicts a gaping wound, any individual who can defend itself gains an immediate advantage. The result? Penis fencing.

As in all fencing, combatants try to hit without being hit; fighting is not to the death but to penetration. Duels can last for an hour, with each contestant striking and lunging, ducking and riposting. It's quite a sight. The creatures look like tiny Persian rugs—flat (obviously) and adorned with intricate and colorful repeating patterns. When they swim, they look like flying carpets. When they fence they look like invisible men dueling under long capes. A duel ends when one animal succeeds in stabbing the other with its penis. This is still not pleasant, but at least the loser put up a fight.

I hope I've convinced you that a bit of slap and tickle isn't necessarily sinister. Indeed, to come back to your case, male seaweed flies gather important information during mating tussles. Discriminating males do not mate with every female they overpower. Instead, they mate only with the most vigorous. The reason, it turns out, is that the most vigorous females are those who are most able to survive. No one knows why the males are so fussy. But one possible reason is that seaweed—which is where females lay eggs—appears on beaches unpredictably. Therefore, the more robust females should have a better chance of living long enough to see seaweed arrive with the tide. So don't be shy. Get in there and beat the girls into submission.

∽

*Dear Dr. Tatiana,*

*I'm a sagebrush cricket, and I've just molted into manhood. While checking out my new manly body, I noticed some teeth on my back. This strikes me as a funny place to have teeth. What are they for?*

*Don't Know Much about Anatomy in the Rockies*

Have you heard of a gin trap? It's a trap with spring-loaded jaws held open in a big toothy grin. When an animal steps on the trigger the jaws slam shut, the teeth seizing the quarry so it has no hope of wriggling free. Trappers once used gin traps to catch bears, wolves, mink, sable, and the like. In eighteenth-century Britain, these traps were even used to catch men. I'm not joking: giant steel gin traps were erected to catch poachers bent on filching game from the estates of aristocratic gentlemen.

Nowadays, happily, gin traps for mink and man alike are largely outlawed. But Mother Nature pays no attention to such niceties. The teeth on your back are a gin trap for catching girls. Here's how it works. In your species, it's traditional for the girl to be on top. When you curve your back upward to link genitalia, the flexion of your back causes the teeth of the gin trap to close on her belly, holding her fast. Once caught, she has to have sex with you whether she wants to or not. That's right. The gin trap enables you to rape her.

Why would you do that? Well, it's an ugly world. Female

sagebrush crickets have a wicked habit of their own: they drink your blood. You've probably noticed that as well as the gin trap, you have another odd bit of anatomy, a pair of soft, fleshy hind wings. You wouldn't get far if you tried to fly with them. No, I'm afraid they seem to have evolved for females to nibble. During sex, she takes a bite or two from your wings, then laps up the blood that oozes forth. Afterward, the blood dries and your hind wings become weird mutilated sculptures. Females naturally prefer virgins because only virgins are intact; after all, who wants someone else's chewed goods? But such picky behavior creates a problem for you males, who are naturally keen to mate more than once. That's why when a female climbs on your back to check the state of your hind wings, you grab her with the trap. If you're a virgin, it doesn't make a difference—she'd have sex with you anyway. But if your hind wings have been chomped, it's the gin trap that makes her stay.

Don't worry, you're not alone in having a device to seize unwilling girls. Look at scorpionflies, insects with long, clear gossamer wings splotched with black. Male scorpionflies have a "notal organ," a clamp on their abdomen that they use to hold females down. As in your case, the notal organ is used in all matings; but again, it only becomes a weapon when a male is trying to hold a female against her will. What determines a female's willingness? Whether or not the male can provide a good meal.

Scorpionflies have an old-fashioned mating system: he pays for supper, she puts out. Because scorpionflies are among the vultures of the insect world—they are scavengers, feeding on insect carrion—a classy male will serve up a nice dead insect. Females readily copulate with males who can offer such a gourmet meal, but dead insects are often scarce. To get one a male may have to resort to stealing from a spider—a dangerous occupation. (Tip: If you're a boy scorpionfly, you'll have a big, bulbous penis. If you're

in a spider's larder and the owner tries to stop you, whack her
with your member and she'll back off. Girls, if you ever find yourselves in the same predicament, your best bet is to head butt the
poor spider.)

If a male cannot obtain an insect by means fair or foul, he uses
his salivary glands to secrete a large gelatinous lump—yummy.
Not as yummy as an insect but not bad. Some males, however, are
not up to secreting the lump and don't fancy burgling a spider.
Consequently, they have nothing to offer females and resort to
force.

In both sagebrush crickets and scorpionflies, rape is the work
of the ultimate loser, the fellow on the edge of society who can
spread his genes in no other way. You see, in these species, it's
better to make love to a willing female than to a furious, struggling, resisting one—males with willing partners copulate for
longer, transfer more sperm, and sire more children. But if you're
a male with nothing to offer, you won't be able to seduce anyone—and coercion is your only option. Indeed, natural selection
discriminates against well-behaved losers. If you've nothing to
offer and you don't resort to force, you won't have any children
and your well-behaved genes will perish when you do.

But you shouldn't think that all rapists are desperadoes, guys
who can't get a girl any other way. Rape has also been reported in
lobsters, fish, turtles, birds, bats, and primates. The identity of
the perpetrators is not always known: in the little brown bat,
anonymous males creep through the vast winter roosts, raping
females (and even males) who are hibernating. But among birds
at least, rapists are typically respectable married men. Take the
white-fronted bee-eater, a small, colorful bird that lives in big
colonies throughout central and east Africa. Males and females
form stable couples, nesting together year after year. But don't let
that fool you. White-fronted bee-eaters are hardly living some

matrimonial idyll. Rape is commonplace. If a female ventures from her nest alone, she will probably be chased by at least one male and perhaps as many as twelve; if they succeed in pinning her to the ground, they all jump on her at once and try to mate with her. In some colonies, females have a one in five chance of being raped in a given year. Yet despite the fact that in a given year many males are bachelors—and under the desperado theory would be the chief suspects—the bachelors are innocent. Almost all would-be rapists are paired with other females in the colony. The lesser snow goose, another bird who lives in colonies, is even worse. In this species, married males routinely attack nesting females. A female left alone for an instant will probably be assaulted by the guy from the nest next door; the usual reason she's on her own is that her husband is off trying to rape someone else. In some colonies, each female is the victim of a rape attempt about once every five days.

As far as anyone can tell, however, this dreadful behavior rarely results in conception. The best estimates suggest that only 1 percent of white-fronted bee-eater chicks are conceived through rape; for the lesser snow goose, that figure is 5 percent. So why aren't these fellows content to look after their wives and children like upstanding members of the community? As in most other birds where males sexually assault females, everyone lives in close proximity. Under these circumstances a male doesn't have to look far for his victims: the cost of attempting rape is minimal. Therefore, the reward—in terms of additional children—needn't be enormous for the behavior to persist.

But how do we know the girls aren't actually asking for it? Sexual coercion is always hard to judge: struggling is not necessarily an indication of reluctance. Many females, though, struggle only when their reluctance is real. Take the American lobster.

Females can mate after they've just molted as well as when their shells are hard. Females about to molt start visiting males, and when they find a fellow they like, they move in with him. No struggling there. Likewise, when a hard-shelled female wants to frolic, she'll present her rear to a fellow after the briefest of preliminaries. But an unwilling female will run from a male's advances. He'll give chase and may even try to haul her from her burrow. Or take scorpionflies. Females fly toward males bearing food and mate while they eat; they fly from empty-handed males. If captured, they fight vigorously to escape, twisting and turning their abdomens to avoid genital contact. When a female bee-eater mates with her husband—as she does once every couple of hours while she's laying eggs—she permits him to feed her an insect, then lifts her tail and holds still while he flutters behind her. But when she's accosted by other males, she flees. If they force her down, she presses her rear firmly against the ground and keeps her tail down, a posture that hinders sex. Moreover, before she leaves her nest, she whistles. If her husband is near, the whistle calls him over so he can escort her. It's a good idea: escorted females are hardly ever assaulted.

All this, together with the small fraction of rape attempts that produce offspring, suggests that resistance is not a ruse to attract male attention. Instead of lying back and thinking of England, most females adopt a "death before dishonor" response to sexual assault. Which raises another important question. Since resistance can result in serious injury or death, why don't females evolve to submit? This is not something that has been studied. But my bet is that there will often be a cost to submitting—one that, on average, outweighs the risk of injury or death. Such a cost can be inferred: remember that in some birds a male who suspects his mate of infidelity will make less effort to feed the chicks, and chicks sometimes starve as a result. In some species of scorpionfly, females live off male efforts to provide food and never have to soil their hands by foraging, an activity that would increase their risk of becoming spider fodder. A female who fails to resist rape attempts would have to forage for herself or go hungry. And we know that in some species females suffer significant costs if mates are foisted on them—by a scientist, for example—rather than being chosen freely. In both the fruit fly *Drosophila melanogaster* and the field cricket *Gryllus bimaculatus*, females assigned a partner have fewer children than those who get to choose.

If male birds, lobsters, scorpionflies, and sagebrush crickets can gain from forcing females to have sex, what about humans? I know. The thought that rape could be natural—by which I mean an intrinsic, evolved part of a man's behavior—is distasteful, even offensive. But to be blunt, it is possible. Evolution does not obey human notions of morality, nor is human morality a reflection of some natural law. The deadly sins would be different if they mirrored evolutionary no-no's. Lust, for one, would be deemed a virtue; chastity would be deplored. In principle, rape could have evolved in humans just as in any other animal: if rapists, on average, had more children than other men, any genes underlying the behavior would spread.

This logic is indisputable. Beyond that, however, it is impossible to say anything definite. Nothing is currently known about the genetics of rape in any animal, let alone humans. But suppose some men do turn out to carry genes that predispose them to rape. Would that somehow make rape acceptable? Of course not. Understanding human evolution and genetics may one day tell us why we are the way we are. But it can tell us nothing about what we would like to become.

಄

Girls, if you want the profile of a typical rapist, I can't give you one. In some societies, they are desperate losers. In others, they are married men. In still others, all males, whether young, old, subordinate, or dominant, use force sometimes. There is no general rule. So what's a girl to do? Here's my Guide to Self-Defense:

1. **Don't** attract attention. Hide or be otherwise inconspicuous.
2. **Don't** leave home alone. Hire an escort or, failing that, stick with other females.
3. **Do** avoid groups of idle males. If they congregate at a place you must go to, try to time your visit to coincide with the arrival of other females.
4. **Do** carry weapons. Males tend to be servile if females are well-armed.

# ·8·
# HELL HATH NO FURY

So the guys fight and brawl with each other while the girls live in peace and harmony? Not bloody likely . . .

*Dear Dr. Tatiana,*

*My name's Jerome, and I'm a moorhen. I'm seriously alarmed by the violent behavior of the girls in my species: they're not babes, they're brutes. At the slightest provocation, they leap in the air and start clawing each other. Why are they so truculent, and how can I stop them from killing each other?*

*Bring Back the Ladies in Norfolk*

Don't worry: in most species, females are far too sensible to kill one another. And when they do fight to the death, it's not usually over a man but over something important—like a house. Even then, a fight to the death is rare. It occasionally happens in thrips—tiny black insects with wings that look like miniature feathers. In certain Australian species, females are armed with massive front limbs and kill each other for their ideal home, a gall on an acacia tree. Queen ants can be belligerent too. In the seed-harvester ant, queens cooperate to found a colony and get it up and running: several queens can establish a colony faster than a queen working alone, and it will be less vulnerable to raids from neighbors. As soon as the colony looks like a going concern, though, the queens are at daggers drawn, tearing one another to pieces in battles for control of the nest. But in general, if you're a girl, the rewards from bumping off rivals are not worth the risk of getting killed yourself.

That doesn't mean that girls don't fight over boys at all. As anyone who's been to a girls' school can tell you, things turn nasty when boys are in short supply. Often the problem is temporary: at the start of the breeding season of the smooth newt, girl newts are hot to trot but boy newts, who have a fixed supply of sperm for the whole season, are initially reserved. The result? Bad manners. Since newts set packets of sperm on the ground, single girls barge in on courting couples to steal the sperm just as the male deposits it. But by the end of the season, most girls have lost interest—having mated, they are busy laying eggs, a slow process because they wrap each one in a leaf—and it's the boys who quarrel over any girls who are still keen. Likewise, in the Australian katydid *Kawanaphila nartee*, an elegant twiglike creature related to crickets and grasshoppers, bachelors are in great demand early in the season. That's because these insects like to eat pollen, and in early spring most flowers haven't bloomed yet. But males woo by secreting a large meal, which they provide along with their genes, so that springtime scarcity of pollen is a double whammy: lots of hungry females eager to mate in order to dine and hardly any males who've eaten enough to be able to make lunch. Under these circumstances, when a male rubs his stumpy wings together to announce he's prepared a meal, several females come jumping and wrestle for the right to mate. Once pollen becomes abundant, however, calm is restored. Being better fed, females are less hungry for sex even as more males are able to join the dating game.

But this is nothing. In some species, the shortage of males is chronic. For *Acraea encedon*, an African butterfly, the deficit is severe. More than 90 percent of the butterflies in some places are female. Why? They are infected with the dread disease *Wolbachia*. *Wolbachia* is a bacterium often found in insects; like a shape-shifting monster, it has different effects on different hosts. For these butterflies, it is King Herod, killing little boys early in their embryonic development. Where *Wolbachia* is common, male *Acraea* are rare—and females gather, chasing any butterfly they see in desperate attempts to find a mate.

The usual reason for a dearth, happily, is more benign. In some species where males help with child care, a given male cannot look after all the eggs or young that one female can produce. Ideally, then, each female would have more than one male at her disposal. But that causes shortages—and fighting. Consider the midwife toad *Alytes muletensis*, a mottled olive and orange creature who lives in gorges on the Mediterranean island of Majorca. In the evenings, males croon love songs from hiding places in the rocks. Females who are ready to lay their eggs will answer one of the singers and then hop around to his home to make his acquaintance. If she likes the looks of him, she'll stroke him on the snout—midwife toad for "Let's do it, baby." At that, he grabs her from behind. She responds by stamping in place, and he starts strumming her genitalia with his toes. They keep this up—punctuated by brief pauses—for perhaps a couple of hours, although heavy petting can go on all night. At last, the female goes into spasms and releases her eggs, each one connected to the one before by a thin gelatinous thread, a kind of poor man's pearls. As the eggs appear, the male wraps his arms around her neck, releases sperm, and starts scissoring his legs like a demented gymnast. The point of this antic soon becomes clear: he's winding the string of eggs around his legs, where it will remain until the eggs are ready to hatch into tadpoles.

Which is why the fighting begins. Once a male has a string of eggs, he's unavailable until he's dumped the ripened eggs into a pool of water. That could take anywhere from nineteen days to two months—exactly how long depends on the weather, as eggs develop more slowly at lower temperatures. But females can produce a clutch every three weeks or so, and they must find a male to look after them; if they cannot, they will lose the clutch. Available males become hot items, and females are not ashamed of trying to steal someone else's guy. Females wrestle with each other or, more often, intrude on a courting couple and hug the male from behind so he can't strum properly—in hopes that the rival female will get fed up with his lousy tap dancing.

This combination—males looking after young and females wanting more than one male each—often leads to blows. That's why I suspect the Darwin frog of having pugnacious females. This little green frog with a pointy nose, who makes its home in the leaf litter of Chilean forests, has an obscure sex life. But we do know a few salient facts. Females lay between thirty and forty eggs. Once fertilized, the eggs sit in the damp earth for about three weeks, at which point the muscular twitchings of the developing embryos stimulate males to start parental care. The form of parental care is peculiar, perhaps unique: the males swallow the eggs and brood the tadpoles in their vocal sacs. After about fifty-two days, Dad opens his mouth and, abracadabra, tiny frogs hop out. Obviously, while a male has his vocal sacs stuffed with tadpoles, he can't sing to attract more mates, so brooding is a big commitment. Unlike the Japanese cardinal fish, a species where males brood fry in their mouths, the Darwin frog doesn't eat his children if he sees a girl sexier than his original mate. But since each male can manage only about fifteen tadpoles, a female needs two or three males per brood. Moreover, every brooding male is out of action for over seven weeks. In such a system, competition for available males must be intense.

Even when bachelors are out in droves, though, girls still scrap and bicker if some fellows are better catches than others. It's the same old story: everyone wants to marry the heir to a fortune, no one wants the poor pimply guy in the corner. This is what's going on with your female moorhens, Jerome. All the girls set their caps at the smallest, fattest fellows; no one wants the big gangly one. If the girls start fighting whenever you approach, you must be ravishingly rotund. Why the fuss over small and fat? In moorhen culture, males do most of the work of sitting on eggs. This may not sound strenuous, but it is. Males fade away as the breeding season wears on. Fat guys can keep eggs warm for longer, and females with fat fellows can produce more clutches in a season than girls with slimmer spouses. Given the difference, it's worth fighting to secure a fat mate. And small? Small males fatten up better. For moorhens, short and dumpy is in.

ᴄᴠ

*Dear Dr. Tatiana,*

*I'm a burying beetle. I met my wife when we worked together at a chipmunk's funeral. It was love at first sight, and after a whirlwind romance I thought I'd found paradise. But now she's turned into a frightful harridan. Nag, nag, nag. I don't get a moment's peace, and whenever I try to relax in the evening by doing*

*headstands, she bites me or knocks me over. What have I done to*
*deserve this, and how can I get her off my back?*

*I Hate the Trouble and Strife in Ontario*

Are you sure you're not trying to have your corpse and eat it too? You know as well as I do that when a male burying beetle stands on his head he exposes the tip of his abdomen and sends sexy scents wafting through the air. I suspect that when you do your headstands you're not chilling out but trying to attract a mistress. Call it a hunch, but that might be why your wife finds your behavior galling.

Look at it from her point of view. The two of you must have struggled for hours to bury that chipmunk. Chipmunks are more than two hundred times heavier than burying beetles. If you were lucky, the body was lying on soft earth so all you had to do was dig soil out from beneath it. But if the ground was hard, you would have had to move the corpse, perhaps over the astonishingly long distance of several meters, to soft ground. Once the body was at last interred and out of reach of ants and blowflies, you had to remove its fur and massage its poor dead flesh into a ball, ready to feed to your babies when they hatch and crawl onto the carcass. Lucky babies! Nothing like regurgitated rotting chipmunk to give you the start you need! Picture it: the grubs sitting in the carcass, rearing up and opening their mouths like baby birds when you or your wife chirrs and bends over to give them their lunch. It's heartwarming to think that these maggoty grubs, who now bear no resemblance to their magnificent parents, will one day sport shiny black wing covers scalloped with red just like yours.

But now you and your headstands threaten this blissful scene. Sure, a mistress would be great for you—you would have more children if you could lure a second female to the carcass. But it would be a disaster for your wife. The presence of another woman and her brats would make it harder for your wife to raise all of hers. This is not just because the two families would have to share the chipmunk meat and there might not be enough to go around. Rather, the mistress would probably murder (and then eat) some of your wife's children. (To be fair, though, your wife wouldn't refrain from chowing down the mistress's kids—it's a burying-beetle-eat-burying-beetle world.)

Females of many species stand to lose if their mate takes an additional lover. Sometimes the mistress does a burying beetle and kills the wife's children. In both the house sparrow and the great reed warbler, for example, a male with two mates will help only the female whose clutch hatches first, so to ensure herself of male assistance, a savvy mistress will smash all the wife's eggs. But things are not always so grim. Often the wife loses out simply because a gallivanting male leaves her with more work than she can handle and she's not able to raise as many offspring. Or she may lose because the brats from a second liaison take food or other valuable resources away from her own children. But whatever the reason, females in many species are not interested in sharing their man—and take steps to prevent it.

As you've discovered, scolding, chivying, and harrying a fellow is one way to make sure he doesn't have time for mischief. A female pied flycatcher who catches her mate singing when he should be working isn't subtle in her disapproval, often interrupting him to make him shut up. The story is familiar: a male pied flycatcher who's trying to impress some floozy will hastily depart if he sees his angry wife bearing down on him.

Rather than raging when they feel insecure, some females mount a charm offensive. Female starlings, for example, get all lovey-dovey and constantly beg for sex if they notice their mate courting other lasses. But whether they shower their mate with kisses or brickbats, females everywhere have the same response to girls they suspect of seducing their husbands: hostility. In the northern harrier, a bird of prey from North America, females intimidate possible rivals and attack them if they are carrying food. Female blue tits dive-bomb rivals, often knocking them out of the air. A female starling, as well as distracting her mate with canoodling, will chase a rival whenever she spots one hanging about and will sing ostentatiously to show the little tart what's what. If she finds her mate strutting his stuff at a second good nest hole, she may resort to filling it up with straw, feathers, and other material to make it look occupied, even though she has a nest of her own. A male starling often ends up chasing his wife back to her own nest to stop her beating on potential mistresses.

But he doesn't have it as bad as Mr. *Lamprologus ocellatus.* This fish lives in Lake Tanganyika, one of the Great Lakes of tropical Africa. A male holds a territory; as for many birds, a crucial feature of his territory is the nest sites that he can offer—in this case, empty snail shells. You might not think so, but there's fierce competition for preowned snail shells. Many creatures find them useful shelters. Hermit crabs famously depend on finding used snail shells: rather than growing shells of their own, they move from empty shell to empty shell as they get bigger. It's not always easy to find a shell of the right size, yet many hermit crabs are finicky, preferring to squeeze into shells that are too small rather than wear a shell of the right size that has a hole in it. And just to show how central used snail shells can be to a local community, some hydroids—simple animals related to jellyfish, corals, and sea anemones—prefer to settle on shells occupied by hermit crabs. How can they tell? Hermit crabs scuttle about faster than the shell's original owner, and baby hydroids are attracted by the motion. It's mutually convenient. Hydroids have powerful venoms that might discourage anyone in the mood for a hermit-crab sandwich. For the same reason, some hermit crabs harvest sea anemones and stick them to their shells. In return, the hydroid (or anemone) gets scraps of food and protection from its own predators, which the crab zealously fends off.

In Lake Tanganyika, at least fifteen species of fish want secondhand snail shells in which to lay their eggs. As males of most species cannot move the shells any distance, they have to make do with whatever they find in the vicinity. But in one species, *Lamprologus callipterus*, males can grow to eleven centimeters (four inches) and can easily nip shells with their mouths and carry them back to their territories. (This species may hold the record for males towering over females: females are less than half as long and fourteen times lighter. They can't grow big because they have to fit inside the shells.) The biggest males hoard shells for females to spawn in. As usual, they have no scruples about where shells come from and often raid one another. So perhaps it's not surprising that in the smaller *Lamprologus ocellatus*, where the male can only move a shell by pushing it, he buries any shells on his territory before swimming about and looking pretty as he waits for females to arrive.

If a girl stops by and they take a shine to one another, he digs up a shell and she takes up residence until she's ready to spawn. (If she fails to spawn within a few days, the male kicks her out— what does she think this is, a homeless shelter?) When she does spawn, she lays her eggs in the shell, gluing them to the inside; he fertilizes them, and she stays in the shell, fanning the eggs to keep them aired and clean. Once the brood hatches, she and her little ones live in the shell until the young are old enough to go out into the world.

Female *Lamprologus ocellatus* hate one another. Whoever moves in first does her best to remain in solitary splendor, chasing all visiting females and otherwise terrorizing them. If a second female decides to settle in the territory anyway, she'll have to live in a shell far from her rival's—and even then, she'll face constant harassment. The male ends up breaking up fights and chasing the aggressor back to her shell; since a pair of females attack each other several times an hour, the peace process takes a lot of his time. It's vital, though. Unsupervised, the females would go on fighting until one of them moved out. So, boys, if you're thinking of cavorting with more than one girl, remember the Chinese symbol for peace. It's one woman under a roof.

〜

Females don't have special weapons and rarely fight to the death over a man—pistols at dawn are not their style. But that shouldn't mislead you into thinking they don't fight over males at all. Females will fight out of:

1. **Desperation**—whenever there aren't enough males to go around. Shortages can happen for any number of reasons but are particularly likely in species where child care is time-consuming and the males do most (or all) of it.

2. **Aspiration**—whenever some males are obviously superior to others, that is, when females mated to the best males have more children than those stuck with lesser fellows.

3. **Possessiveness**—in species where males and females form pairs, females go to great lengths to prevent their partner from taking other lovers. To remain the one and only, females attack possible rivals—and for good measure, distract or hound the male who strays.

So if you hear it said that there's a grand, harmonious sisterhood, you're probably hearing propaganda. In most species, it's not "all for one and one for all." It's every girl for herself.

# APHRODISIACS, LOVE POTIONS, AND OTHER RECIPES FROM CUPID'S KITCHEN

---

**W**hat do homosexuality, new species, and love potions have in common? If you look closely, you may find that they are all outcomes of the battle of the sexes at its most fundamental.

---

*Dear Dr. Tatiana,*

*I think I've made a dreadful mistake. I've just lost my virginity to a fellow who recently escaped from a local fruit fly laboratory. He's bigger and badder than wild flies—and he says he's put a spell on me so that I'll never be able to have sex with anyone else. Do you think that's possible, or is he all bluff?*

*Afraid I've Been Bewitched in Santa Barbara*

In general, it's a good idea for wild flies to steer clear of escapees. All manner of experiments go on behind the walls of fruit fly laboratories, and time inside can do strange things to a fly. As for laboratories breeding lovers with supernatural powers, powers that could never evolve in the wild, I fear your seducer may indeed have been telling the truth. How did he get that way? To understand, we're going to need a more intimate look at the battle of the sexes.

Remember: while a female may gain from mating with several males, each of her lovers will do better if she mates with no one but him. Whenever this conflict of interest occurs, it ignites an evolutionary battle that is fought on two different fronts. On the first front, each of a female's lovers tries to thwart the lovers of her past and future while avoiding being thwarted himself. The prickly penis is one device we've brushed against before, and high sperm counts should sound familiar; but as always, there are other possibilities. For example, a male may use chemicals to disable his predecessor's sperm. Or he may make his own sperm hard to remove.

Meanwhile, the second front is male versus female: he evolves tricks to manipulate and control her, and she evolves to resist. One obvious and common ploy is for the male to try to stop the female from taking subsequent lovers: regular readers of these columns will recall chastity belts and the fanatical guarding of a female from rivals. But again, that's only the beginning. There are many other, more exotic—and in some cases, more sinister— ways in which males of various species attempt to bend females to their will. To give just a few examples: males may attempt to increase the number of eggs a female lays right after sex. Or increase the number of their sperm that she stores. Or deliver drugs that switch off her sex drive—an invisible, chemical chastity belt. Or daub her with an "antiaphrodisiac," a chemical that makes her stink, so that other males will find her repulsive. In short, the arsenal of potent weapons is enormous. With so much scope for innovation, the battle will inevitably unfold in different directions in different species—or even in different populations of the same species.

Seminal fluid—the liquid that sperm, ahem, come in—is typically a complex brew, containing many chemicals that alter female behavior. In the Australian field cricket, for example, chemicals that accompany the sperm stimulate egg production: inject a virgin female with the right ingredient and she will start laying eggs even though she has not copulated. The seminal fluid of the housefly contains at least twelve active proteins, some of which act like a drug, binding to receptors in the female's brain and turning her off sex. Fruit fly seminal fluid is even more complex, containing more than eighty proteins. The function of most of these is still a mystery. We do know, however, that some play a role in disabling a previous male's sperm while others make sperm harder to remove. Yet another protein has antiaphrodisiac properties. And another, a small molecule known as sex peptide, inspires the female to lay perhaps fifty eggs and induces aggression toward males. If any male approaches within a day of her getting the molecule, the female will give him a good kicking.

Hermaphrodites get in on the act, too. Consider *Helix aspersa*, the ubiquitous garden snail. Mating individuals sling arrows of outrageous fortune at each other: each has a love dart that is sharp and pointed and that, if fired, pierces the lover's skin. As it does so, it delivers a gob of mucus. The mucus contains a substance that alters the female part of the partner's body, widening the passage to the sperm-storage chamber and closing the entrance to the sperm-digestion chamber, thereby increasing the chance that arriving sperm will be kept for future use rather than sent to the sperm-digestion chamber and destroyed.

The fact that males have powerful effects on females does not by itself, however, demonstrate that there's a conflict of interest.

Take red deer. During the breeding season, stags spend most of their time roaring. A stag's roar is a long, low rumble; each roar is a single exhalation of breath. It used to be thought that roaring was simply a kind of ritual fighting, a contest to see who is stronger. But a stag with a harem roars much more than he needs to if intimidating the opposition is the only aim. He'll typically roar at least twice a minute all day and all night—that's nearly three thousand roars every twenty-four hours, not counting extra roaring during shows of strength. Small wonder that after a couple of weeks he's exhausted. It's worth it, though. For females, roars are an aphrodisiac: females exposed to vigorous roaring come into heat sooner than females who are not. But this effect is not necessarily malevolent, in his interests but not hers. Female red deer who conceive earlier in the breeding season give birth earlier the following spring and are more likely to have their calves survive. Responding to roars may help females by increasing the odds they'll get pregnant early in the season.

In order to demonstrate that these love philters and aphrodisiacs are nefarious, then, you need to show not only that they effect a manipulation but that the manipulation is being resisted—that the female is fighting back. One way to show this is to arrange matings between individuals from far-flung populations: if females are evolving to resist male manipulation, you'd expect them to be more resistant to their local fellows than to fellows from far away, with whom they have not been directly coevolving. Among houseflies, for example, if you capture flies from distant places—Sweden and the United States, say—and have them mate with each other, females tend to be less susceptible to the monogamy potion made by males from their hometown than they are to the potion made by the foreigners.

A second, more powerful way to reveal the battle is to do experiments that change the underlying conflict. And I'm sure you won't be surprised to hear that such experiments have already been carried out on fruit flies.

In one experiment, male and female interests were forcibly aligned. How do you forcibly align their interests? Each generation, you imprison couples together for the whole of their lives, so that neither male nor female has a chance to mate with anyone else. Under such circumstances, you'd predict that males would evolve to be less manipulative of females. Correspondingly, females would have nothing to struggle against, and thus their ability to fight back should deteriorate. Sure enough, this is what happens. After eighty-four generations of enforced monogamy, male fruit flies allowed back into the meat market of the regular mating arena were much less effectual at preventing females from mating with other males. Likewise, the females had become more vulnerable to male manipulation: in comparison with flies who'd come from eighty-four generations of promiscuous mating, females accustomed to monogamy laid far more eggs for their first lover and consequently took much longer to agree to mate again.

The forcible alignment of male and female interests occasionally occurs in nature, with couples imprisoned together forever. In some species of shrimp, for example, couples are trapped within a Venus's flower basket, a glass sponge from the deep sea. Each sponge is shaped like a cornucopia with a lid, thanks to its elaborate—and beautiful—latticed skeleton. Shrimp arrive as babies and crawl inside; once they start to grow, they cannot get out again. Apparently, they are only ever found in pairs—I suspect they kill any latecomers. No one knows whether they've evolved a life of perfect sexual harmony; from the rules of war outlined above, however, that is indeed what I would predict.

To return to the fruit flies in the trenches, though, a second experiment produced an even more compelling result. Usually, conflict is difficult to show unequivocally because males and females are evolving in lockstep. In this experiment, one of the combatants, the female, was not allowed to fight back: by dint of fancy genetic techniques, females, but not males, were prevented from evolving. This gave males a stationary target to adapt to instead of the usual moving one. In such a situation, the females can no longer hold the males in check—male evolution will no longer be constrained by the female ability to fight back—and without this opposition, males should continue to evolve in a direction beneficial to males.

The results were clear. In less than forty generations, the males had, I'm afraid, become supermales, completely redrawing the battlefield. Females were much more likely to be seduced by these fellows, failing to reject their advances even when under the influence of a regular male's sex peptide. At the same time, the supermales were better at inducing monogamy, and even if a girl did eventually mate with a normal fly, the supermales were less likely to have their sperm displaced. In a nastier development, some of the supermales seemed more likely to cause their mates to die, perhaps because their seminal fluid had become poisonous. So if your lover escaped from an experiment like this one, I can well believe that he breached all of your natural defenses.

Which inspires an arresting question: if human males could evolve for forty generations while females were held still, would they become a breed of dangerous but irresistible superlovers? I wonder.

∽

*Aloha, Dr. Tatiana,*

*I'm afraid of becoming a has-been of a rock-boring sea urchin. In my species, sperm are frightful fashion victims: they constantly change their outside coats. The rumor surfing over the reef is that this is because eggs are snobs and only the trendiest sperm are allowed to penetrate them. How can I find out what this year's fashion will be, and can I get my sperm engineered to match?*

*Desperate to Be à la Mode in Hawaii*

As so often with rumors, the one you've heard is a garbled mix of fact and fantasy. Let me walk you through what's really going on. You know, of course, that the first step in fertilization is for the sperm to attach to the egg. In your species—indeed, among sea urchins generally—the sperm attaches to the egg using a protein called bindin. When you hear talk of sperm changing their coats, what is actually meant is that they are changing their bindin; and it's true that in the grand scheme of things bindin is changing rapidly. But there's no need to get het up—rapid is relative. Your brand of bindin will be in fashion thousands of years after your shell has smashed against the reef.

You're probably wondering why, if a change in bindin isn't imminent, everyone is making a fuss about it. Well, the reason is that your brand of bindin is fundamental to your identity as a rock-boring sea urchin. I'm not exaggerating. Consider this: like rock-boring sea urchins, oblong sea urchins—your near relations—shed their eggs and sperm into the water. In principle, then, one of your sperm might encounter one of these foreign eggs. But if it does, nothing will happen: you are powerless to fertilize it. This is because your bindin is the wrong shape for attaching to an egg from an oblong sea urchin.

So what? So a lot. Reproduction is central to the concept of a species: the working definition of a species is a group of organisms that can interbreed. Any mechanism that prevents interbreeding can thus generate new species. For example, lengthy physical separation between two groups often generates new species, as each group goes its own way. This is why islands and lakes (which are, after all, islands of water surrounded by land) tend to be rich in unusual flora and fauna.

New species can also arise, however, through basic reproductive incompatibilities—such as egg and sperm not recognizing each other. Continuing with your case, a change to the bindin-egg interaction is enough, by itself, to generate a new species of sea urchin. To see how, suppose that the only difference between male rock-boring sea urchins is the type of bindin that they produce. To keep things simple, let's say the bindin comes in just two types, A and B. And suppose that the only difference between female rock-boring sea urchins is the affinity of their eggs for one type of bindin or the other. If the affinity is so strong that some eggs can be fertilized only by sperm carrying type A bindin, whereas the rest of the eggs can be fertilized only by sperm carrying type B, then rock-boring sea urchins would be two species, not one—even though individuals are identical in all other respects.

Matters have not yet reached such a pass. They are heading in that direction, though. Male rock-boring sea urchins do produce bindin of distinctly different types, and the eggs do have affinities for one type or another. The affinities are not—yet—exclusive. But they will become so if the different bindin-egg interactions continue to diverge. Indeed, such a process seems to have happened before. If you compare yourself with the oblong sea urchin, you'll see that the biggest genetic differences between the two of you concern those governing the egg-sperm interaction. In other ways, you hardly differ at all. So you see, the rapid evolution of bindin really does have ramifications for your species.

Curiously enough, in a number of other organisms the proteins involved in reproduction—and therefore in making sperm, eggs, the molecules contained in seminal fluid, and so on—are also evolving fast. Very fast. In mammals, for example, two proteins found on the surface of the egg, both of which interact with sperm during fertilization, are evolving at a gallop. In fruit flies, a protein that influences whether a female's subsequent lovers will be able to remove a fellow's sperm, is likewise evolving quickly. But the champion is lysin, a protein that determines whether an abalone sperm can enter an abalone egg: it's evolving as much as twenty-five times faster than gamma interferon, a protein important in the immune systems of mammals and one of the fastest-evolving mammalian proteins yet discovered.

This gives us two intriguing facts. First fact: many proteins involved in reproduction are evolving unusually rapidly. Second fact: changes to such proteins can be instrumental in the origin of new species. Putting the two together: if only we knew *why* proteins involved in reproduction evolve so rapidly, we could gain great insight into the forces underlying the origin of species.

One reason that proteins involved in reproduction evolve so fast may be the battle of the sexes. It's certainly an attractive theory—males and females locked into the battle of the sexes are, as we've seen, engaged in a fast-moving evolutionary arms race. Moreover, in insect groups where most females mate with more than one male—the precondition for war—new species arise at least four times faster than in groups where most females mate only once.

This finding is an encouraging start, and bodes well for the theory. It is too early, however, to leap to any definite conclusions. To show that the battle of the sexes is driving the origin of new species, we would ideally want to show that rapid evolution in a given male reproductive protein is fueling the rapid evolution of a given female reproductive protein, and vice versa. To give a hypothetical example, suppose that sea urchin sperm are evolving to penetrate the egg ever faster. From the egg's point of view, this is bad news: faster penetration may mean that more than one sperm succeeds in entering the egg. But the egg doesn't "want" to be entered by more than one sperm, since in your species more than one sperm prevents the embryo from developing. Therefore, an egg that resists being penetrated too rapidly could have an advantage. To date, though, we have only one example of a fast-evolving protein where both sides of the story are known, and the results are something of a cautionary tale: it's more of a chase than a war.

As I said a moment ago, lysin, the protein that determines whether an abalone sperm can enter an abalone egg, is evolving at record speed. Tantalizingly, abalone are also splitting into new species at a startling rate. So let's look at what's going on with the egg. Like sea urchins, abalone release eggs and sperm into the water by the thousands. Each abalone egg is enveloped in a fibrous substance called vitelline. In order to enter an egg, the sperm must break a hole in the vitelline; the breaking is done by lysin. Thus the lysin carried by the sperm must first be able to attach to the envelope of the egg. Changes to the envelope, therefore, may make it harder for lysin to attach. If lysin cannot attach properly, the sperm carrying it cannot get inside. Therefore any changes to the envelope should quickly be matched by changes to the lysin carried by the sperm.

It turns out that lysin is indeed evolving in response to the egg. The egg, however, is not evolving in response to lysin. Instead, the envelope changes by accident, and lysin races to keep up. This is puzzling. Lysin evolves at record speeds while the egg drifts nonchalantly along? The key to the mystery lies in the nature of the egg's envelope. It is a complex entity, a mosaic of several components. The particular bit of it that lysin attaches to is a giant molecule called VERL—vitelline envelope receptor for lysin. It is made up of one unit repeated twenty-eight times.

Repeated units often create genetic problems. For example, suppose a gene contains a stretch where the same letter, or pair of letters, of the genetic alphabet is repeated several times in a row. When the time comes to copy the gene—for example, when making eggs or sperm—the genetic copying machinery of the cell slides along the DNA molecule, making a replica of the gene as it does so. All goes well until the copying machinery comes to the repeats. Then, all too often, the machinery slips, loses its place, and accidentally puts in too many repeats or too few. It's as if you were trying to copy the number 7878787878787878787 without punctuation marks and without being able to go back and check it. And yes, this can wreak havoc. Several human genetic diseases— the dementia known as Huntington's disease, for example—are caused by accidental changes in the number of repeats.

In the case of abalone's VERL, the unit that is repeated is big—rather than starting again every second letter, it starts again every 460th letter—so slippage isn't the problem. But something else peculiar can happen. When a repeated unit is large, a mutation that occurs in one unit may spread gradually through the other units through a passive process known as concerted evolution. Exactly why this happens is not well understood. But it can have serious consequences. Suppose that a mutation occurs in one of VERL's twenty-eight units. And suppose the mutation is so severe in its effects that the unit is unrecognizable to lysin. From the egg's point of view, this is irrelevant. The other twenty-seven units of each VERL molecule will still function, so the mutation won't affect the egg's chances of being fertilized. Initially, this won't make any difference from the sperm's point of view either since the sperm will be able to find an unmutated unit to attach to. But if the mutation starts spreading through the other units, then any male who produces a version of lysin that can recognize the new unit will start to have an advantage. And I'm afraid it is this passive process that seems to be driving lysin's extraordinarily rapid evolution—and causing abalone to split into new species. Rather than a battle, abalone is engaged in an evolutionary pursuit.

This result doesn't invalidate the idea that the battle of the sexes may be an important cause of the origin of species. It simply means that we have to be careful not to jump to conclusions until we have more data. As for sea urchins, I'm afraid that little is known about the egg side of matters, so I can't say yet whether you're a sex warrior, a skirt chaser, or something else altogether. If I find out, I'll let you know.

∽

*Dear Dr. Tatiana,*

*My son cuts a fine figure of a manatee, and I'm very proud of him. But there's one problem. He keeps kissing other males. What can I do to straighten him out?*

*Don't Want No Homo in the Florida Keys*

It's not your son who needs straightening out. It's you. Some homosexual activity is common for animals of all kinds. Look at the bonobo, a sensual creature also known as the pygmy chimpanzee (which is odd, as it's no smaller than the regular chimpanzee). Bonobos like sex, and female bonobos like sex with each other. One lies on her back, another climbs on top, and they rub genitalia. Among Adélie penguins, one of the smaller penguins in Antarctica, the males, like most birds, have no penis. But that doesn't stop a bit of gay jiggery-pokery. In one recorded incident, two males bowed to each other as they would to a female, then one lay down on his front, raised his bill and tail like any coquettish girl penguin, and the other copulated with him, ejaculating into his genital tract. They then swapped roles. Or look at dolphins. The bottle-nosed dolphin is catholic in its choice of sex partners. Males are frequently sighted copulating with turtles (they insert their penises into the soft tissues at the back of their victim's shell), with sharks, and even with eels. Eels? Yes, when a dolphin's penis is erect, it has a hook on the end—and many a male will use it to hook a writhing, struggling eel. So it should be no surprise that males also copulate with each other, inserting their penises into each other's genital slits. The Amazon River dolphin, or boto, sometimes goes further, penetrating another dolphin's blowhole. So I wouldn't worry about a little kissing.

Why do they do this? Maybe they like it. In the stump-tailed macaque, a gregarious Asian monkey with a (guess what?) stumpy tail, females achieve orgasm through female-female mounting just as they do from heterosexual copulation. Or maybe it serves a social function. Among baboons, homosexual behavior between males seems to facilitate teamwork. Males who mount each other and fondle each other's genitals are more likely to work together when fighting against other males. Or maybe it has an antisocial function. Among razorbills, northern black-and-white seabirds that resemble puffins, males mount each other as a display of aggression. Male razorbills do not like being mounted; they never solicit it, nor do they cooperate if it happens. Instead, they either fight back or run away. Males who are mounted a lot become intimidated and give up their efforts to compete for mates.

Or maybe homosexual activity springs from desperation. That's the most logical interpretation of a copulation witnessed between two octopuses, not just of the same sex (both male) but of different species. These octopuses live twenty-five hundred meters (eight thousand feet) under the sea and presumably meet other octopuses of any sort rather rarely. Almost nothing is known about these particular octopuses: the tryst was the first time that a member of either species had made itself known to science. In general, however, male octopuses do not live long once they mature sexually so, if mates are scarce, perhaps no potential partner should be passed up. And in several species of gull, females are more likely to form pairs with each other when males are scarce. Female couples build a nest together, defend the nest together, and help each other incubate their eggs. But although the girls mount each other and display to each other as they would to male partners, neither adopts a "male" role. In case you're wondering, the eggs are fertilized by males paired to other females in the colony. Female couples are less successful, hatching fewer eggs and rearing fewer chicks than conventional pairs. It's better than nothing, however: without some kind of partner, they wouldn't be able to raise any offspring at all.

In most species, though, homosexuality doesn't lead—even indirectly—to reproduction. So it might seem odd that homosexuality should persist. Indeed, among humans, this is taken as proof that homosexuality cannot be genetic. But it can. In fact, from an evolutionary point of view, homosexual behavior is only perplexing if every one of three conditions is met. First, the behavior must have a genetic basis. If there is no genetic basis, homosexuality cannot be subject to natural selection. Second, it must be exclusive. That is, at least some individuals engaging in homosexual behavior must be confirmed bachelors or confirmed spinsters who never attempt to breed. If they breed, it's no mystery why the genes persist in the population. Third, such individuals must make up a significant proportion of the population. If the behavior is rare, it can be dismissed as a fluke. But if it is common, then there must be something that keeps the genes for homosexuality present at an appreciable frequency *even though the individuals expressing the genes never reproduce.*

So, are the three conditions met? With regard to the first, I should say at once that we know little about the genetic basis of homosexual behavior in any organism, let alone manatees. In humans, the search for genes involved in homosexuality has so far been inconclusive. Some studies find a link, others do not. In fruit flies, however, mutations in a number of genes affect a fly's homosexual proclivities. Some mutations cause males to court males and females indiscriminately. Some cause males to court only males: put several males together, and they will pursue one another in a ring. Bizarrely, one mutation causes males to engage in bisexual behavior only when the lights are on. Extinguish the lights and you extinguish their libidos: these flies don't like sex in the dark. Of course, it is difficult to extrapolate from fruit flies to humans or manatees. But it seems likely—indeed, I would bet on it—that sexual orientation in mammals will turn out to have some genetic basis.

As to the second condition, we know even less about the extent to which an individual's homosexual behavior is exclusive. Among humans today, particularly those in the West, some individuals are undoubtedly exclusively homosexual for their whole lives; the extent to which this has always been true, however, is unknown. Social pressures to marry or leave an heir may have meant that most individuals with strong homosexual preferences reproduced anyway. And in other animals—well, we can't begin to say. Manatees are not known to develop exclusive homosexual preferences. Among captive Japanese macaques, though, males and females sometimes fight each other to have sex with a popular female. (This monkey reaps the benefits of a long tail: a female often rubs her clitoris with her tail.) In captive rhesus monkeys, males sometimes prefer to have anal sex with each other rather than to copulate with females. Whether, given the choice, such individuals maintain their preferences to the exclusion of breeding is unknown.

And the final condition—the incidence of the behavior? We have no idea for any animal. Even among humans, measuring the incidence of exclusive homosexuality is tricky.

But for the sake of argument, let's assume that, in some species at least, all three conditions are met: homosexuality is genetic, exclusive, and common. What can stop the genes' disappearance? Or, to put it more formally, how can the genes be maintained in the population?

The traditional explanation is that the genes can be maintained if homosexual individuals act to increase the reproductive success of their relations. This is because related individuals share a certain proportion of their genes. Identical twins, for example, have all genes the same. A child shares half its genes with each parent. Full brothers and sisters who are not identical twins share, on average, half their genes. (I say "on average" because each gets half his or her genes from each parent, but the halves will be drawn at random.) Similarly, first cousins share, on average, one eighth of their genes. And so on.

This means that you don't necessarily have to reproduce to spread your genes. Instead, you can devote yourself to helping your relations spread theirs. Such a calculus often explains apparently altruistic behavior, such as that of the ants, bees, and wasps who slave away for the good of the colony and never reproduce themselves. But there is no evidence that homosexuality amounts to an indirect way of spreading genes; in species of birds and mammals where young animals help their parents raise the next brood, there's no evidence that the helpers are prone to homosexuality. On the contrary, they typically go on to breed themselves in the following season.

A variant of this idea draws inspiration from another aspect of animal social life. Among animals that live in highly organized groups, such as termites, snapping shrimp, wolves, and naked mole rats, only a few individuals breed. Everyone else is a nonreproductive worker. At least in the case of wolves and naked mole rats, the reason workers don't breed is not because they are physiologically incapable of reproducing but because their reproduction is suppressed by the dominant pair. It is possible, in principle, that homosexuality could evolve as a type of reproductive suppression, with homosexual individuals a nonreproductive caste akin to the warriors in a termite colony. The argument would be that for those organisms who must live in groups to survive, natural selection acts on groups rather than on individuals. If groups with homosexual individuals do better, in evolutionary terms, than groups without them, then the trait will be maintained. This could work if, for example, homosexuals engaged in activities that were good for the whole group, such as hunting or fighting to defend the group from raids by neighbors. But again, there is no evidence that this explains homosexual behavior in any species. In addition, the idea has a theoretical weakness. Unless groups are also family units—as they are in termites, snapping shrimp, wolves, and naked mole rats—the forces favoring reproductive suppression are likely to be weak. Unrelated individuals, after all, have little reason to cooperate in having their reproduction suppressed.

Alternatively, genes for homosexuality could be maintained if the genes themselves are favored by natural selection. This could happen in two ways. The first is known as heterozygote advantage. Suppose a given gene has two possible forms. Since you get two copies of the gene, one from each parent, you could have two copies of the first form, one copy of each, or two copies of the second form. Geneticists say there is a heterozygote advantage when having one copy of each is better than having two copies of either one. The textbook example is resistance to malaria in humans. The molecule that carries oxygen around the body in the bloodstream, transported in red blood cells, is hemoglobin. A variant of the hemoglobin gene, known as sickle cell, causes the hemoglobin molecule to take the wrong shape once it has released its oxygen, causing the red blood cell to collapse. If you have two copies of the sickle cell gene, you'll have severe anemia—and without intensive medical treatment, you won't live long. But if you have only one copy of the sickle cell version of the gene, you are resistant to malaria. The only drawback—and it is a severe one—is that two parents, each immune to malaria, risk one child in four's dying of sickle cell anemia.

With respect to homosexuality, the idea would be, again, that a given gene has two forms. Two copies of the first form, and you're an average heterosexual. Two copies of the second form, and you are homosexual—from a genetic viewpoint, sterile. But one copy of each, and you would have some enormous advantage—for example, you might be a highly fertile heterosexual. To me, this seems an unlikely explanation for homosexuality, for three reasons. First, few examples of heterozygote advantage have been identified. Second, in the malaria example, being a heterozygote prevents you from being dead—which, from an evolutionary point of view, offsets the reproductive cost of having some of your children die. But in the case of homosexuality, it is hard to imagine how, with respect to natural selection, the benefit from being a heterozygote could be large enough to compensate for a certain number of your offspring being sterile. Third, it is unlikely that homosexuality is controlled by a single gene.

I can imagine one other way that genes for homosexuality could be directly favored by natural selection: through an insidious and fundamental form of conflict between the sexes. Namely, if there are genes that produce homosexual behavior in members of one sex but confer great reproductive success on members of the other. From a theoretical point of view this could work. We know that genes that are beneficial in one sex can spread, even if they are detrimental in the other. Under some circumstances such genes can spread even if the benefit to one sex is small while the detriment to the other is large. Thus, suppose a gene causes exclusive homosexuality in males but unusual fecundity when it occurs in females. The gene could persist—and perhaps even spread—despite its reproductive disadvantage in males. The more genes are involved, the more plausible the mechanism becomes. It would also then follow that male homosexuality and female homosexuality are due to different sets of genes.

But forget theory. What about data? Although the results do not pertain explicitly to homosexuality, there is experimental evidence that sets of genes beneficial in one sex can be highly detrimental in the other. As usual, the study animal was the fruit fly *Drosophila melanogaster*. Using the latest genetic wizardry, scientists were able to arrange for eggs from different females to be fertilized by males bearing genetically identical sperm. When the eggs hatched, all the fruit flies, both males and females, thus had one set of genes in common and one set of genes different. Comparing these flies with ones drawn at random from the regular population, the scientists were able to evaluate whether possessing a given set of genes is beneficial, first with respect to succeeding as a larva (measured by whether you make it to adulthood), and then with respect to succeeding as an adult (measured by how many offspring you have).

The scientists tested forty different sets of genes. The results were striking. First, the sets of genes that help you succeed as a larva are beneficial whether you're male or female. This makes sense: the demands of larval life are the same regardless of your sex. For adult flies, however, the sets of genes that confer success on males turned out to be different from those that confer success on females. Indeed, genes beneficial in one sex are detrimental in the other—and the greater the benefit, the greater the corresponding detriment.

These results suggest that the mechanism I outlined to explain homosexuality is at least plausible. Beyond that, however, the possibility that genes beneficial in one sex are detrimental in the other suggests that whether you are a fruit fly, a human, or a manatee, the battle of the sexes is inescapable, insoluble, an eternal war.

～

Girls, when you jilt one beau and go to bed with the next, pause for a moment to consider the countless ramifications, the evolutionary mischief that may result from your lust. As you fall asleep in his arms, picture the various weapons at work and imagine their gradual refinement over the millennia. Ponder whether you and your men are inadvertently playing a part in the slow creation of new species and pray that your genes are well suited to your sex. For the only creatures who can hope to be exempted from this war are the rarest of the rare: the truly monogamous. We turn to them in the next chapter.

# TILL DEATH DO US PART

Real happily-ever-after, till-death-do-us-part romance is almost unheard of in nature. Who has it—and under what circumstances does it evolve?

*Dear Dr. Tatiana,*

*My husband and I have been faithfully married for years, and we are shocked by what we read in your columns. As black vultures, we engage in none of the revolting practices you advocate so regularly, and we don't think anyone else should either. We suggest you champion fidelity or shut up.*

*Crusading for Family Values in Louisiana*

Remember: revulsion is in the belly of the beholder. And if I may say so, gorging on carrion is considered revolting in some circles. You complain, however, that I do not often urge or even discuss monogamy. I'll be frank. True monogamy is rare. So rare that it is one of the most deviant behaviors in biology.

Let me remind you of a few facts. Before the 1980s, more than 90 percent of bird species were believed to be monogamous at least for the duration of the breeding season, and many couples were thought to mate for life. But better genetic techniques and the spread of paternity testing exploded that idea. The halos rolled away like hoops—and not just from birds. Closer inspection of other famously monogamous groups, such as gibbons, revealed that they, too, are not the saints they seemed to be. Nowadays, if animals live in pairs, they are said to be socially monogamous—a term that makes no assumptions about their sex lives. And the discovery of true genetic monogamy is headline news.

So who's making headlines? Genetic testing has revealed a scattering of species, like yours, where couples appear to stay faithful. I say "appear" because few studies have covered more than one breeding season or more than a few dozen families; additional information could change the picture. But with that caveat, genetic data support the claims of the jackdaw, a bird that looks like a small crow with a gray skullcap, to be truly monogamous. The chinstrap penguin appears beyond reproach, as does the long-eared owl. Kirk's dik-dik, one of the smallest African antelopes, seems to have a good case. The California mouse has an excellent reputation. Some termites are devoted lovers. Yet apart from their unusual loyalty, these groups have nothing obvious in common. So let's take a step back and consider when true monogamy, the real thing, boy and girl together forever, might evolve as a stable strategy for all members of a population.

The short answer is that monogamy will evolve as a strategy for all members of a population only when it is in the best interest of both males and females. That is, monogamy will be stable only when members of loyal partnerships have more surviving children than fickle creatures do. Lasting fidelity is rare because it's seldom in the interest of one of the parties, never mind both.

The usual explanation for monogamy in a species is that a female cannot rear offspring alone. In this scenario, which I call the Good Wife Theory of Monogamy, opportunities for males to philander are predicted to be curtailed by female virtue: would-be philanderers will be unable to find partners because each female will be obsessively faithful to her mate for fear of losing his help with the kids. In other words, monogamy is a female plot forced on males.

The theory is unlikely, however, to apply widely. In the first place, it assumes that the male would gallivant if he could—and that being faithful is not actually in his interest. But as we shall see, that assumption is often wrong: sexual fidelity can be in the male's interest. Second, monogamy can evolve even if the male does not help the female raise the children. Look at Kirk's dik-dik. The buck does nothing to help his mate; he just trots after her wherever she goes. He's not even any use at detecting predators that could endanger his calf. The African hawk eagle can pick off a baby dik-dik but not an adult, yet only the female reacts to the eagle's cries.

The third problem with the Good Wife Theory is that a female's need for male help does not necessarily guarantee her fidelity. Take the fat-tailed dwarf lemur. No bigger than a squirrel, this tiny nocturnal primate lives on the island of Madagascar. For the seven months of the dry season, fat-tailed dwarf lemurs hibernate, curled up in tree holes. Like other hibernators, they fatten up before turning out the lights; they store the fat in the base of their tails. In the days before DNA tests, they would have been held up as models of monogamy. Males and females live together in pairs: they snooze together during the summer days, and the male helps with the kiddies, baby-sitting while the female is resting or gathering food. Females cannot raise offspring on their own. Yet genetic testing shows that infidelity is rife, with males often helping to raise children who are not theirs. Of course, this doesn't mean there's never an association between sexual fidelity and parents cooperating to look after the young. But as we shall see, when fidelity and cooperation go together, monogamy is typically the cause and not—as the Good Wife Theory would have it—the consequence.

Let's look now at other, more plausible ways monogamy can evolve. Suppose females are few and far between. Then, a male who finds one may be better off staying with her and keeping rivals away. Indeed, if females are scattered across the landscape, either of two factors may make long-term monogamy particularly attractive for a male. The first is if it is risky to leave—if, say, finding another girl would entail a long or dangerous journey. The greater the risk, the greater the incentive to remain. I call this the Danger Theory of Monogamy. The second factor is if the time between a female's broods is short. There may be no point in setting off on a journey, whatever the risk, if your girl will soon be ready to breed again—this is the Pop-'Em-Out Theory of Monogamy.

The mantis shrimp *Lysiosquilla sulcata*, an animal that ambushes passing fish by spearing them on its fearsome front limbs, is a fine example of how danger can encourage lasting love. Although the genetics of this case have not been investigated, circumstantial evidence for monogamy is strong. Animals pair up as adolescents, then each couple digs a burrow in a sandy spot on the seafloor. Since these mantis shrimp hunt from the burrow's concealed entrance—one of the pair lies in wait, only its eyestalks poking out—individuals need never step outside. Indeed, leaving the burrow is almost certainly a death sentence. Whereas mantis shrimp of most species are covered by armor plates, *Lysiosquilla sulcata* are not. Instead, they have soft bodies—great for burrow dwelling but hopeless against predators. And even supposing an individual managed to roam the seabed without being eaten, building a new burrow would be impossible. To prevent the burrows from collapsing, these animals stabilize the sand with a mucus they produce; as adults, they lose the ability to produce the mucus, so any mantis shrimp wanting out of a relationship cannot simply leave the burrow and build another. Thus, regardless of whether domestic bliss has soured into loathing, it's better to stay than to go.

Now let's turn to another example. Why should a girl tolerate some oaf hanging about? Well, one way for boys to ingratiate themselves is to give her a hand. If a female gains significantly from having a male around, she may be less likely to throw him out. For example, he could help her defend her territory. Or he could help her with child care. The male Djungarian hamster makes himself useful in just this way. Djungarian hamsters live in arid parts of Mongolia. They forage for seeds, which they stuff into pouches in their cheeks; on arriving back at the burrow, they unload their cargo by pushing on the pouches with their forepaws so that seeds stream forth. The male is such an attentive father that he plays the midwife for the births of his pups (the only male mammal so far known to do this as a matter of routine), helping them emerge from the birth canal, opening their airways so they can breathe, and licking them clean. For good measure, he eats the placenta, a delight usually reserved for the female. Consistent with the Pop-'Em-Out Theory, female Djungarian hamsters live apart from one another: their ranges do not overlap. And they are prolific breeders. In a year, they can produce eighteen litters of between one and nine pups each. In contrast, their close relative the Siberian hamster breeds for only a few months of the year and exhibits neither paternal care nor monogamy.

Are there any other ways monogamy can evolve? Certainly. Monogamous organisms are often aggressive toward any animal who's not their partner; it's usually thought that they are aggressive because they are monogamous. But sometimes they may be monogamous because they are aggressive. Suppose that individuals who are aggressive toward members of their own sex have more surviving offspring than more amiable creatures. Then monogamy may arise as a side effect of that aggression. This is the Sociopath Theory of Monogamy. As a possible example, consider banded shrimp. These animals look like candy, with long white antennae like spun sugar, their bodies and claws white but encircled with big red bands. They typically live in twos, feeding each other and standing guard while their partner molts. Each member of the pair, however, is ferociously aggressive toward shrimp of its own sex and will fight such individuals to the death. I know people like that.

Alternatively, monogamy may arise when cheating or desertion by either partner results in total reproductive failure for both. This is the Mutually Assured Destruction, or MAD, Theory of Monogamy. One group thought to be MAD lovers are hornbills—birds found throughout Africa and Asia. Hornbills come in various shapes and sizes—there are around forty-five species—but are easily distinguished from other birds by their large, curved bills, which are betopped with a horny casque. Again, even without genetic evidence, the case for monogamy is strong. In many of these species, males and females are completely dependent on each other for the survival of the brood. At the start of the breeding cycle, she climbs into the nest hole and either she or her mate seals the entrance so that only her beak can poke out. The male then feeds her while she lays her eggs and incubates them; once the chicks have hatched, he brings food for the whole family. It's a demanding job: in the wreathed hornbill, for example, the female may remain inside the nest for as long as 137 days. She molts her flight feathers shortly after she starts to lay eggs, so if he should disappear, all the nestlings and the female will die. Contrary to what one might imagine, this arrangement has not evolved as a trick to prevent the female from cheating; rather, it seems to be a defense against nest predators.

But not all monogamous organisms kill members of their own sex whenever they meet one, sequester themselves or their partners in tree trunks for weeks on end, live scattered across dangerous landscapes, or produce wagonloads of offspring. Jackdaws, for example, often live in colonies, where opportunities for seducing the neighbor's spouse are abundant. Yet they don't seem to take advantage. The reasons for their self-restraint aren't clear, but I'd guess there's an element of MADness. Baby jackdaws are difficult to raise successfully: in bad years, 80 percent of couples fail to fledge young. This suggests that both parents must devote all their energies to the nesting effort; anyone wasting time philandering will definitely see the nest fail. There may also be an element of experience. In species that live for many years—and jackdaws do—working with a familiar partner can greatly enhance reproductive success. Among Bewick's swans, for example, the longer a couple has been together, the better they'll be at raising cygnets.

And black vultures? Well, black vultures do tend to nest apart from one another and are fiercely territorial when it comes to protecting the nest. But that can't account for their fidelity, as they still have plenty of opportunity to encounter potential lovers. While one of the pair is sitting on the eggs, the other will go off foraging—and black vultures often meet up at carcasses or even at roosting sites. But it seems that these birds have a social convention that helps individuals stay faithful. Apparently, black vultures insist that sex be conducted in the privacy of the nest and won't tolerate lewd behavior in public. If a young bird who doesn't know better tries to get laid at a roost, the poor creature will be roundly attacked by the other vultures in the vicinity. Who'd have thought black vultures would be so prudish?

୬

*Dear Dr. Tatiana,*

*Like, what's the deal? I'm a sleek young California mouse and am so in heat. But, like, the guy next door won't pay any attention to me? Even though his wife is totally old and ugly? No way is he still in love with her. How come he's, like, such a dull dude?*

*So Bummed Out in Berkeley*

However old and ugly their wives, male California mice are not to be lured away from them. Once one of these dudes, as you call them, has a partner, he won't cheat on her even if he finds himself locked up with a virgin in heat. So if you want a man, you'll have to find one who's single. Bad luck.

But perhaps you're wondering what accounts for your neighbor's saintly behavior. Well, I guess that infidelity is just not in his nature, that he's a breed of anti-Lothario. It's easy to imagine how he could have got that way. These days, California mouse couples are truly monogamous, living together faithfully until one of the partners dies. If you look at their near relatives, however, you'll find all sorts of lascivious goings-on, so it's safe to assume that these virtuous rodents are descended from promiscuous ancestors, ancestors who squeaked their bawdy way from one bed to the next.

If you've been reading my columns, you'll know that a population can only evolve from promiscuity to monogamy if incorruptible couples consistently have more surviving children than libertines do. If this happens, and if monogamy has a genetic basis, then genes associated with monogamy will spread. Eventually, everyone in the population will have these genes. (Note, however, that the genes influencing fidelity need not be the same in males and females.) So it's possible—indeed, I'd say probable—that your mouse dude has a powerful genetic predisposition toward monogamy.

With respect to California mice, I can't say much more than that. In the case of your rival for the title of Ultravirtuous Rodent, the prairie vole, though, scientists are starting to figure out the genetic basis of monogamy. So let's take a brief look at what's been discovered about how it works in males.

When a boy prairie vole meets a girl prairie vole and they decide to become an item, they consummate their relationship by copulating anywhere from fifteen to thirty times in twenty-four hours. From this point on, the lovers grow greatly attached, endlessly cuddling and grooming each other, the picture of mushy affection. And that's not all. Before losing his virginity, a male is a peaceful sort of chap, not prone to picking fights; but after his first night of passion his personality changes. Now if he sees any prairie vole—male or female—who is not his partner, he'll attack vigorously. What accounts for his transformation?

Sex. For a male prairie vole, the sex act causes the release of vasopressin, a hormone that binds to special receptors (the vasopressin $V_{1a}$ receptors) in the brain and alters his behavior. We know that vasopressin is responsible because, if you inject a mating male with a chemical that blocks vasopressin from binding to the receptors, he will behave as though he has not had sex. Conversely, if you inject vasopressin into a virgin male, *he* will behave as though he has.

Strong stuff. But before any girls out there start dosing their boyfriends with vasopressin, I should say that this trick won't work on everyone. For although vasopressin is found in all mammals, it has different effects in different species. As proof of this, consider the montane vole, a close relative of the prairie vole. Montane voles mate promiscuously, they do not form stable couples, lovers rarely sit companionably together or groom each other, and sex induces neither affection nor aggression. Inject a male montane vole with vasopressin and he won't pick fights. He'll groom himself.

These different responses to vasopressin appear to be due to the way vasopressin $V_{1a}$ receptors are distributed through the brain—a distribution that differs considerably between the two species. Intriguingly, it's quite straightforward to create the prairie vole's distribution. All you need is a mouse embryo and the gene that contains the instructions for building the prairie vole version of the receptor. Combine the two—which gives you a mouse carrying the prairie vole gene—and you'll get an adult mouse with vasopressin $V_{1a}$ receptors distributed in the prairie vole pattern. Give this mouse a shot of vasopressin and it will start to behave like a prairie vole in love.

These results give us a glimpse of the underlying mechanism of monogamy; the full story will probably turn out to be much more complicated. Moreover, since monogamy has evolved independently in different species, the mechanisms may well differ from one monogamous species to the next. All the same, if sex also turns out to have potent hormonal effects in other monogamous species, this would explain why many of them—from the Indian crested porcupine to the wood roach *Cryptocercus punctulatus*—regularly engage in sex that, owing to its timing, cannot possibly result in reproduction. (It's particularly amazing that porcupines have far more sex than necessary for reproducing—porcupine sex is potentially pretty prickly.) Irrespective of such matters, however, it is important to remember that understanding the genetic mechanisms of monogamy is quite separate from understanding why organisms have evolved to be monogamous in the first place.

What, if anything, does all this mean for humans? For now, nothing is known about the genetics of human monogamy, but I'd like to indulge in some speculation.

Taken as a species, humans cannot be described as exclusively monogamous. Divorce rates and extramarital affairs attest to that. Also, nonidentical twins occasionally have different fathers (although how often this happens is unknown since it's not usually detected unless the fathers come from different ethnic groups). But humans are not wildly promiscuous, either. Some individuals are faithful to one partner for their whole lives; few have or admit to having thousands of sex partners. Moreover, a couple of indices put humans as a whole on the monogamous end of the scale. First, consider physical differences between males and females. In monogamous species, males and females tend to be roughly the same size; in species where a few big males hold harems, males tend to be enormously bigger than females. Remember the southern elephant seal, where adult males are typically twice the length of adult females—and may be as much as ten times heavier? In the gorilla, adult males are typically twice the weight of adult females. In contrast, human males tend to be only slightly bigger than females, and there is considerable overlap, with some women bigger than some men.

Next, consider physical differences between human males and other male apes. As you know, testicle size is usually associated with the risk of sperm competition. Males that are at low risk of sperm competition—either because they are good at defending a harem or because they are paired with a faithful female—generally have testicles that are small in relation to their body size. Males that are at high risk of sperm competition—either because they pursue a strategy of seducing the partners of other males or because most females mate promiscuously—generally have enormous testicles in relation to their body size. Male gorillas are at low risk of sperm competition and have tiny testicles. Male chimpanzees are at high risk of sperm competition and have gigantic testicles. In comparison, human males have medium-sized testicles, suggesting a low to moderate risk of sperm competition. This fact, plus the relatively small size differences between men and women, is exactly what would be predicted for a mostly monogamous species.

What about the rate of infidelity as measured by genetic paternity testing? This is always rumored to be extremely high among humans—30 percent or more. But in fact, surprisingly few studies have even looked at it: despite an extensive trawl through the scientific literature, I could find only a handful. Of these, most showed a low rate of infidelity—3 percent or less; the highest value I found was 11.8 percent. These results need to be treated with more caution in humans than in other species: contraception and abortion enable humans to avoid having a child during an extramarital affair. In the past, however, these would have been less of a factor, and there are ingenious ways to get at historical infidelity. For example, in England, children usually get their last name from their father. Boys get something else from their father: their Y chromosome. Thus, if all living males bearing a particular last name are the direct descendants of one man, they should all have the same genetic markers on their Y chromosomes. In the absence of infidelity (or adoption), last names and Y chromosomes should match up. One study analyzed the Y chromosomes of men called Sykes, a name that first appears in written records about seven hundred years ago. It turns out that almost all the Sykeses investigated did indeed have the same markers on their Y chromosome, suggesting that most living Sykeses have the same distant ancestor. The rate at which females married to Sykeses were unfaithful (or adopted sons) over the period of seven hundred years is estimated to be 1.3 percent per generation. But maybe there's something odd about Sykeses. It would be great to have more studies on other names.

Supposing that humans as a whole are best described as mostly monogamous, we are left with two questions. First, what forces may have led to mostly monogamous humans? There are several possibilities—including, perhaps, cultural pressures. Beyond concluding that more monogamous people must generally have had higher reproductive success than less monogamous people, we cannot say much. Second, do individual humans—just like individual crickets and fruit flies—differ in their genetic predisposition toward monogamy? To put it another way, social mores aside, do some people have a much easier time complying with their wedding vows than others have? It is tempting to imagine that once we know more about the genetics of human behavior we will find not only that different men have different proclivities for monogamy but that a given proclivity goes along with a set of other traits—as, you may recall, it does in the California singing fish and among dung beetles. Perhaps it will turn out, for example, that men with large testicles (anticipating a high risk of sperm competition) are prone to seducing other men's wives and have difficulty forming lasting bonds whereas men with small testicles (anticipating a low risk of sperm competition) are prone to sexual fidelity and jealousy and turn all lovey-dovey after sex. But for now, this is all conjecture . . .

                               ∽

For most boys and girls, wedding rings are made of fool's gold—real, true love is precious and rare, the confluence of bizarre biological forces. Several factors may contribute to monogamy, but you'll find that true love works best when it is absolutely MAD.

# PART III

# ARE MEN NECESSARY?
# USUALLY, BUT NOT ALWAYS.

Male and female form a fundamental, immutable dichotomy, the front and back of a single coin, her yin to his yang.

Or do they?

Actually, he is less essential than she is. That's right, a species can do without him but not without her. Some species reduce the number of males to the bare minimum. Some get rid of males altogether—and simply don't bother with sex. What's more, there's nothing immutable about the two sexes, nothing preordained about sex roles. Some species even transcend yin and yang: they have sexes, sure, but nothing as boring as females and males. Who? When? Why? How? Read on . . .

# ·11·
# THE FORNICATIONS
# OF KINGS

Within human societies, incest is traditionally reserved for royalty. Other organisms, however, are not so elitist. And guess what? Regular practitioners of incest dispense with males almost entirely. So when is it acceptable—or even desirable—for sex to be a family matter?

*Dear Dr. Tatiana,*

*Something terrible has happened. I'm a male mite of the species* Acarophenax mahunkai—*the scourge of the lesser mealworm beetle. This morning I was, as usual, making love with one of my sisters when my mother's belly burst. All my sisters wandered off, leaving me alone inside Mom's corpse. Is this punishment for messing with my sisters? What will happen to me now?*

*Aghast in Arkansas*

Bad news: you're finished. All you can do is stagger about on your eight stumpy legs in the hopes of finding a stray sister so you can mate once more before you die.

Life's not fair. Not only have you had your day, but you're not even really the scourge of the lesser mealworm beetle. That's your sisters. You're just an accessory to their crimes. Let me explain. The girls in your family suck eggs—the eggs of the lesser mealworm beetle, to be precise. When a mite sucks an egg, her belly swells up to twenty times its usual size and she becomes a huge balloon with a tiny head and legs—the mite version of a caricature of a grotesquely fat man. Her children—as many as fifty—develop and copulate inside her, then she bursts. The newly emerged female mites seek out any lesser mealworm beetles who have succeeded in hatching and stow away on the beetles' undersides like so many tiny scabs. The female beetle unwittingly carries this deadly cargo with her when she goes to lay her eggs. (Can female mites distinguish between male and female beetles? I would guess they can, but nobody knows.) Meanwhile, you male mites rarely manage to leave your mother's body—and die almost before you've lived.

Will you go to hell for having screwed your sisters? Don't worry about that. Whether or not hell exists, incest is not intrinsically bad. If you could flick through *Who's Who in Nature*, you'd find a multitude of organisms who, like you, habitually practice close incest without ill effect. True, it's not advisable for everyone: among humans, for example, the children of brother-sister or father-daughter matings are likely to be sickly or deformed. But this is not divine retribution for monstrous sin. It's a simple consequence of genetics.

Problems with incest are due to recessive genes. What's a recessive gene? Elementary, dear mite. Humans and most other sexually reproducing organisms are "diploid": they receive two copies of each gene, one from their mother and one from their father. If the two copies are different, how they interact to influence a trait, such as eye color, varies—but the outcome can be simple, with one copy overriding the other. The overriding copy is known as dominant, the overridden copy as recessive. Thus, the effects of a recessive gene don't show unless an individual has inherited two copies of it. Such an inheritance can be deadly. Recessive genes are more likely not to work properly, so a double whack can be disastrous, resulting in immediate death or debilitating disease.

When a recessive gene is rare, however, it can persist unseen because most of the individuals who harbor it will have only one copy. Herein lies the danger of incest. Because family members are genetically more similar to one another than they are to strangers, sex in the family raises the odds of uniting two copies of a harmful recessive gene. The closer the kinship of the lovers, the more genes they will have in common—and the greater the risk that harmful recessives will be expressed in their children. Let me give you an example. A recessive gene lurking in one person in a hundred is seven times more likely to meet its twin in the child of a first-cousin marriage than in a marriage between two strangers plucked at random from a crowd. For the child of siblings, the risk is twenty-five times greater. And that's when the recessive gene is quite common. If the gene is rare—say, lurking in one person in ten million—the risk for the child of siblings is 2,500,000 times greater than for a child of strangers.

When "inbreeding depression"—the reduced vigor of inbred children compared with outbred children—is severe, close incest is unlikely to catch on. The reason is simple. Anyone who prefers sex with strangers over sex with family members will have more and healthier offspring. If this preference has a genetic basis, genes for outbreeding will spread. Which may explain why humans and some other mammals avoid mating with those who were around them when they were kids. Children reared on Israeli kibbutzim provide a good example: in the golden age of the kibbutz, children were raised in communal children's homes rather than in small family groups. As adults they reported, often with regret, having no erotic interest in those who'd been their childhood companions—and out of 2,769 marriages of kibbutz children, none took place between former playmates.

Given all this, you're probably wondering how—and why—anyone ever gets involved with their siblings, children, or parents. Well, several factors can reduce the severity of inbreeding depression or otherwise tip the balance in favor of incest. Broadly speaking, regular practitioners of close incest fall into two groups. In one, you have hermaphrodites—committing the most intimate incest of all, self-fertilization. (Selfing isn't as much fun as it sounds: it doesn't usually involve copulating with yourself. Instead, eggs and sperm shuffle around inside you. But there are exceptions. The earthworm *Dendrobaena rubida* folds on itself, pressing its male parts against its female parts.) The other group of habitual inbreeders is made up of organisms, such as you, my mitey friend, that routinely bed their nearest and dearest. Joining you in these bacchanalian festivals of the family are the (human) royalty of Hawaii, ancient Egypt, and Inca Peru, many other mites, a few pinworms, and a profusion of insects. Otherwise, recurrent, close incest is rare. So what makes these groups special?

Consider hermaphrodites first. Roughly 80 percent of flowering plants are hermaphrodites, either making perfect flowers—flowers with both male and female parts—or making separate male and female flowers simultaneously. Of these, perhaps three-quarters self-fertilize once in a while, but far fewer self-fertilize exclusively. Among animal hermaphrodites, much less is known about the proclivity for selfing, but what is clear is that many don't—indeed, can't. Upon reflection, this is not surprising: no one is more related to you than you are to yourself, so the off-spring of a selfing hermaphrodite are even more vulnerable to inbreeding depression than the fruits of a sibling union. But any child is better than none, and a hermaphrodite who cannot find a mate should eventually switch to selfing, even if the inbreeding depression is appreciable—a strategy I call "emergency selfing." The white-lipped land snail, for example, prefers outcrossing, but if no mate has appeared after about a year, a lone snail will give up the wait and start selfing. For hermaphrodites, then, the incli-nation to self-fertilize should depend on the balance between inbreeding depression and the opportunities for mating with others.

Let's turn now to nonhermaphrodites. Among the Hawaiians, the ancient Egyptians, and the Incas, brother-sister mating was believed to be a mark of divinity. According to Egyptian mythol-ogy, Geb, the god of the earth, married his sister, Nut, the god-dess of the sky, while the Incas, who also worshiped incestuous deities, believed that the sun married his sister, the moon. The royal families of both societies claimed divine ancestry—and mimicked their ancestors' behavior. In old Hawaii, if a chief of the highest rank married a full sister, their son was considered holy and anyone who came into his presence had to lie down. (The unfortunate fellow was apparently confined to his house during the day so that ordinary people wouldn't have to keep dropping everything and falling flat on their faces to worship him.) If the big chief merely married his half-sister, however, the child was considered less awesome: people had only to sit before him.

Mythology notwithstanding, it's worth asking why such practices began and then why they didn't die out as imbeciles and cripples ascended the thrones. The historical record is silent on the latter point: we don't know the extent to which inbreeding depression made itself felt. In all three societies, however, kings had a number of other wives; in the event the sibling union proved infertile or produced an individual unfit to rule, another heir could presumably be selected. As to the origin of the custom, one idea is that sibling mating is a natural consequence of rigid social stratification. In stratified societies, few men are suitable matches for the highest-ranking women. After all, cousin marriages (accompanied by a lesser but still detectable degree of inbreeding depression, such as hemophilia and the Hapsburg lip) were common among European royalty, at least in part for this reason. If women are never permitted to marry men of lesser rank than themselves, there will be times when the only mates available to the top women are their brothers.

Delusions of divinity play no part, however, in the evolution of close incest among insects, pinworms, and mites. So what accounts for their enthusiastic embrace of it? Crucially, many insects and mites and all pinworms have few recessive genes, which means negligible inbreeding depression. But why do these critters get away with fewer recessive genes than the rest of us? The answer lies in the marvelous genetic systems they employ.

Two particular genetic systems—each of which has evolved several different times—are extremely efficient at purging recessive genes. The first, known as haplodiploidy, is the more common. In this system, females, like humans, are diploid: they receive two copies of each gene, one from each parent. Males, in contrast, are "haploid": they hatch from an unfertilized egg and thus receive only one copy of each gene, from their mother. In other words, males have no father and females don't need to mate to produce sons. That's right: a boy's mother may be a virgin.

This permits all sorts of jolly debauchery. Take the button beetle *Coccotrypes dactyliperda*. This creature lives in grottoes that it hollows out of date stones (or, indeed, buttons—yes, I do mean the buttons on clothes). Brother and sister button beetles can mate with each other right after hatching—but that's just the beginning. On arriving at a new home, a female who failed to mate with one of her brothers in the date stone of her birth digs out a grotto and then lays a small clutch of unfertilized eggs. These develop into males. She mates with the first to hatch and then eats him and his brothers before laying a large brood of daughters—and perhaps one or two more sons for her girls to mate with.

Worse: the wasp, *Scleroderma immigrans*. The female paralyzes beetle larvae with repeated stings before drinking their blood. Next, she plasters their bodies with eggs so that her growing children can also enjoy the bloody feast. In this species, a mother not only mates with her son but also goes on to mate with a grandson produced by a daughter from the first incestuous liaison. It puts Oedipus in the shade.

The second genetic system is less versatile, less common, but much more weird. Called paternal genome elimination, it is practiced by various mites and a smattering of insects. Here, males arise from fertilized eggs as they would in humans. But then— this is the weird part—early in the embryo's development, the cellular machinery inactivates or destroys the father's genes. The result is that once again, to all intents and purposes, males have only one set of genes.

So you see, in both systems harmful recessive genes never have a chance to accumulate. Since males have only one copy of each gene, recessive genes are never hidden behind healthy ones: any flaws are immediately apparent and thus exposed at once to the full fury of natural selection. This means that males bearing harmful recessive genes will die. On the other hand, inbreeding depression is unlikely in the event of incest. When your mitey ancestors first turned to their siblings for comfort, recessive genes didn't stand in their way.

∽

*Dear Dr. Tatiana,*

*I'm a true armyworm moth, and I've gone deaf in one ear. I've read this is from having too much sex. Trouble is, I'm (sob) still a virgin. So what's happening to me?*

*Piqued in Darien*

Be assured, you have nothing to worry about. It's just that your inner ear is now hosting a torrid, incestuous orgy. Remember the rhyme you learned as a caterpillar?

A moth who can't hear
At all in one ear
Is probably quite
A home to a mite.

*Dichrocheles* infest
While the moth is at rest
An unlucky event—
Mites never pay rent.

Yet once they're on board
They are all in accord
'Cause they've learned to perfection
Through natural selection
(Or heard from an oracle)
To invade just one auricle.

For a moth who's stone-deaf
To the ultrasound clef
Is lunch for a bat.
No doubt about that.

What happened is that one evening when you stopped to sip nectar from a flower, a mite scrambled up your tongue as if it were a ladder. When she reached your face, she crawled through the tangle of your scales and hairs to the outer caverns of your ears; after inspecting both, she chose one and crept inside. Then she stepped up to the delicate membrane—the tympanic membrane—that screens off the inner ear from the outer ear, and she pierced it. In doing so, she destroyed forever your ability to hear with that ear.

After settling in and perhaps taking a light supper of—I'm afraid—your blood, she started to lay her eggs, about eighty in all. A couple of days later, the eggs hatched, the little larval mites wriggling backward out of their eggshells. First to emerge were the males of the brood; then came all their sisters. The males grew up faster than their sisters, prepared one of the innermost galleries of your ear as a bedchamber, carried their sister-brides thence, and even helped them out of their old skins as they finished their final molts into adulthood.

You needn't feel self-conscious about having your ear infested by mites—such things happen in nature all the time. Army ants—the ones that sweep through rain forests killing everything in their path—have one species of mite that lives on their antennae and another that lives on their feet. Hummingbirds get mites in their nostrils when they drink at flowers. The mites don't cause the birds to lose their sense of smell, since they are just hitching a lift between flowers. But they are still a nuisance: nectar thieves, they can slurp up as much as half of the nectar a flower produces.

Humans play host to the (mostly harmless) mite *Demodex folliculorum*, which lives in eyelash follicles, as well as to *Demodex brevis*, which occupies the sebaceous glands. Fruit bats have mites on their eyeballs. Birds have mites inside the quills of their flight feathers.

But coming back to the orgy in your ear, there is one thing I'd like to draw your attention to. Namely, that the orgy embodies what some would consider paradise. Of the eighty or so eggs laid by the mother mite, one or perhaps two will have hatched out males; the rest will have hatched out females.

This deserves notice because, in general, dramatic deviations from a sex ratio of one to one are rare. Males and females of most species are born in roughly equal numbers. The reason for this balance was first explained by Ronald Fisher—the same fellow who suggested females might mate with attractive males in order to have sexy sons. In essence, his argument is one of supply and demand. Suppose girls were more common than boys. Then, any parent with a predisposition to have sons would have more grandchildren than a parent with a predisposition to have daughters because boys would be scarcer and more likely to find mates. The gene for sons would spread—and as it did, the sex ratio would become less skewed. Since the same argument applies if the initial skew were toward more boys, the only stable situation is a sex ratio of one to one, and any deviation should thus be quickly and automatically corrected.

Humans may provide an instance of such a correction. During wars, large numbers of men are killed, skewing the sex ratio toward women. Therefore, one might imagine that a sex-ratio adjustment would occur in response to wars. Consistent with this, a significantly larger proportion of boys were born in the immediate aftermath of each world war than before the outbreak of hostilities. (I should stress that the mechanism for this is unknown and the finding may be a coincidence rather than a demonstration of Fisher's principle in action. But it is provocative nonetheless.)

A pronounced departure from a one-to-one sex ratio, therefore, indicates that something unusual is afoot. The "something" can be sinister: numerous parasites meddle with their host's sex ratio in order to increase their own transmission. In the common woodlouse, for instance, microbes transmitted only through eggs act to convert genetic males into females. The wood lemming—a tiny, stocky rodent with a furry tail that lives in the bogs and forests of northern Europe, Siberia, and Mongolia—has a maverick chromosome that skews the sex ratio toward girls. As a result, the typical wood lemming population is more than 70 percent female.

More benignly, a highly skewed sex ratio is one of the trademarks of close incest. In *Syringophiloidus minor*, the mite that lives in the quills of the house sparrow, each female lays twelve eggs, only one of which will hatch into a male. Likewise, *Acarophenax mahunkai*, our aghast correspondent from Arkansas, would most likely have had his fifty sisters to himself. In short, among the unrepentantly incestuous, males are produced with extreme thrift.

The reason for this was discovered by Bill Hamilton, one of the most original and important evolutionary biologists of the twentieth century. He pointed out that among inbreeders a female will often arrive at a new home—be it a coffee bean or a date stone, your ear or my eyelash follicle—that no one else may ever attempt to colonize. Under these circumstances, the argument for a one-to-one sex ratio breaks down. When a female is isolated, her reproductive success depends only on how many daughters she has: she will not gain any additional grandchildren from her son's seducing the daughters of other females, because there are no such individuals to be seduced. Therefore, the original matriarch should produce just enough sons to inseminate her own daughters: any more would be a waste of time and energy.

In general, then, habitual incest allows for all kinds of savings with respect to the structures involved in male function. For hermaphrodites, it means a minimal investment in male apparatus. For example, in the hermaphroditic mussel *Utterbackia imbecillis*, a parasite that lives on the gills of freshwater fishes, the proportion of its body devoted to sperm production diminishes as incest—in this case, selfing—increases. Selfing plants dispense with showy flowers: they have no need for extravagant displays to attract pollinators. Similarly, incestuous females do not gain by creating huge, macho sons—all that matters is that the boys live long enough to shag their sisters. Sure enough, the sons of inbreeders tend to be precocious runty fellows with short life expectancies. Often, they do not feed during their brief glimpse of life on earth; many don't even have mouths. Soon after their ecstatic orgies are done, these males die, usually without having left their natal bean, or quill, or ear. I'm afraid, dear moth, that once your tenants have disembarked onto scented blossoms to wait for the passing of a fresh host, their brothers' rotting bodies will remain, a leperous ghost colony in the inner porches of your damaged ear.

You do have something to thank them for, though. The first mite apparently leaves some sort of trail, for if a second or a third mite should get on board to start a family, the new arrival will go to the ear that's already occupied. Indeed, if your deaf ear is already overflowing with occupants, the mites of this species will not invade your intact ear—just as the old rhyme says. They would rather get off again and wait for a new moth than deafen you to the ultrasound squeaks of bats. This makes sense: if you die, they die. But their having evolved such an unerring response suggests that multiple boardings are not uncommon. And that, in turn, suggests that incest is not always the only choice.

If other females begin to show up in your ear, sons have a chance to mate with females who are not their sisters—and the advantage of producing more males starts to increase. Thus, the more females who colonize a particular spot, the more sons each one should produce and the more balanced the sex ratio should become. Take *Nasonia vitripennis*, a tiny wasp that lays her eggs in blowfly pupae. When the female finds a pupa, she drills through the wall and injects a venom that kills and preserves the developing blowfly. She lays eggs, of which perhaps 10 percent will be males, and then heads off to seek new pupae. If, however, she arrives at a pupa that is already occupied, she will adjust her brood so that she has more sons.

How does she do this? Well, *Nasonia vitripennis* is one of those species where males hatch from unfertilized eggs. This system readily allows an exact control of the sex ratio: each mother can determine how many sons and daughters she will have by how many eggs she fertilizes. The mites inside your ear, however, have the other genetic system that aids and abets incest, the one where males develop from fertilized eggs but the father's genes are promptly discarded from male embryos. At first glance, it seems unlikely that females in this system would have the ability to fine-tune the sex ratio of their offspring according to circumstance. And yet—amazingly—they seem to be able to. Experiments on *Typhlodromus occidentalis*, a mite that has paternal genome elimination, show that females produce more sons in the presence of other females. Does such a shift also occur in your ear? No one knows, but I would guess it does.

༾

*Dear Dr. Tatiana,*

*Like any decent, upstanding mangrove fish—that's* Rivulus
marmoratus *to you—I've always fertilized my own eggs. But
this evening I came home to the burrow I share with a friendly
great land crab and I found a stranger had moved in. He claims
that he's a mangrove fish as well—but that he's a real man, not a
bit of both like me. He says he wants us to do all sorts of terrible
things together. It sounds like fun. But will it do me any harm?*

*Gagging for It in Florida*

Normally, I'd say go for it. But in your case, we need to consider
the situation carefully. You mangrove fish have some bizarre cus-
toms, one of which is your habit of clambering out of the water to
spend time on land, locomoting with flips, wriggles, and jumps.
This explains how you get into the burrows of the great land
crab—the entrance to a crab's home is usually reachable only
from terra firma. Even more remarkable, you can survive more
than two months out of water—quite a feat for a fish.

To a geneticist, however, what makes you unusual is the fact
that you're the only vertebrate—the only animal with a back-
bone—known to self-fertilize. Indeed, hermaphrodite mangrove
fish are incapable of spawning with one another; they can only
spawn with individuals who are pure male. When males are rare,
as they are in Florida, a population of mangrove fish consists of
hermaphrodites who have been selfing for generations.

With that in mind, let's examine the risk you could run if you
mate with a male. You've heard of inbreeding depression? Well,
ironically enough, there's also something called "outbreeding
depression." The idea is that sometimes unions between distantly
related individuals produce offspring who have a poor chance of

surviving or reproducing—a worse chance, in fact, than the off-spring of more closely related parents.

In principle, outbreeding depression can occur for two reasons. First, the partners may have genes that cannot act effectively in concert. The most obvious, most prevalent, and least interesting example of this comes about when an organism tries to mate with a member of a different species. Such bestial couplings are generally infertile; after all, species are defined as groups that cannot interbreed. But when species have recently separated, offspring may still result from cross-species fornications. The coupling of a mare and a jackass famously yields a mule—but note that outbreeding depression still strikes, since the mule is sterile.

Less extreme outbreeding depression may indicate that populations are in the process of diverging into separate species. In pink salmon, for instance, individuals have a fixed two-year life cycle. Thus, a given river may have two salmon populations separated in time: even-year salmon and odd-year salmon. Under ordinary circumstances, ne'er the twain shall meet. When they do meet—as eggs and sperm in the test tubes of a laboratory—they beget offspring that are viable but that don't survive as well as purebreds.

The second potential cause of outbreeding depression is the external environment. Suppose individuals have evolved features that help them cope with particular local conditions. Then, mating with an individual from somewhere else, who is therefore not well suited to prevailing conditions, may break up favorable gene combinations. Consider the soapberry bug. This creature makes a nuisance of itself by feeding on seeds of the soapberry tree. To get at the seeds, it pierces the fruits with its beakish mouthparts—which are the perfect length for reaching the seeds. Recently, however, the soapberry bug has started feeding on seeds of the round-podded golden rain tree. In this plant, the seeds are buried more deeply within the fruit—and the ideal beak for a soapberry bug needs to be much longer. As a result, soapberry bugs have evolved to specialize on one tree or the other: short-beaked soapberry bugs live on soapberry trees, and long-beaked soapberry bugs live on round-podded golden rain trees. Sex between members of the two populations could disrupt beak length, which could make it impossible for the progeny of such crosses to feed.

Before you conclude that you're damned if you inbreed and damned if you outbreed, I should say that documented cases of outbreeding depression within a species are far fewer than documented cases of inbreeding depression. True, outbreeding depression has been studied less. But that's not the only reason for the difference. In many species, inbreeding depression drives the evolution of elaborate mechanisms that stop organisms from mating with their relations. In contrast, mechanisms to avoid outbreeding are virtually unknown. Thus, to the extent that it occurs, outbreeding depression is probably not a cause of mating patterns but a consequence—and a trivial one, at that. Let me give you an example. Outbreeding depression is most likely when two mates come from populations that don't normally interact. After all, pink salmon from different cycles don't mate, because they don't meet. In the case of some plants, outbreeding depression may arise as a consequence of the activities of pollinators. Bees often fly fixed distances between flowers—and it's not uncommon for crosses between plants more than one bee flight away to show outbreeding depression. Therefore, my guess is that for most of us outbreeding depression will turn out to be of little concern.

But mangrove fish are not most of us. Outbreeding depression could well be an important force for organisms—like yourself —

that have a history of selfing. Thus, we must at least consider whether your offspring will suffer outbreeding depression if you depart from tradition and mate with a male. I can't give you a definite answer, but what we know so far suggests that you can go ahead and whoop it up with your new friend if you want to. Here's the evidence.

*Rivulus marmoratus* is found in mangrove swamps up and down the Atlantic coast of tropical America. Males are vanishingly rare: in most populations, they occur only sporadically. In Honduras, however, males make up around 2 percent of the population, while in Belize they are for some reason quite common (around 25 percent). In both places, hermaphrodites regularly allow themselves to be seduced by males. Although the fertility of the resulting offspring has not been measured, outcrossing has no obviously harmful effects. Outcrossed fish are the same size and shape as inbred fish and have no detectable oddities or deformities. So if you're bored with yourself—say yes to sex!

ఞ

To sum up, incest is not the prerogative of kings, pharaohs, and chieftains; more often, it better befits the humble residents of the king's eyelash follicle than his majesty himself. But whether your blood is regal blue or common red, I advise you not to frolic with family if you carry harmful recessive genes. For the time being, alas, there's no way to know in advance whether you have such genes. I therefore provide some rules of thumb.

## THE PLEBEIAN'S GUIDE TO INCEST

1. If you feel a strong aversion to erotic activities with your relations or if you're a hermaphrodite and you possess features

that make selfing difficult, it's probably a sign that incest is not for you. But even when carrying recessive genes, hermaphrodites should still be prepared to self—if possible—in emergencies.

2. If you are not a hermaphrodite, incest is best if you come from a species where males have only one set of genes. If you're not a member of such a species, I urge you to avoid sex with your nearest and dearest. As a last resort, you can kiss your cousins—but stay away from siblings, parents, and grandparents.

3. Finally, if you decide to make a career of incest, make sure to slash your investment in males to a minimum. There's no point wasting energy making armies of big, strong sons if these fellows will just mate with their sisters or their mothers and then die. That's right: the best mama's boys are small and weedy.

# · 12 ·
# EVE'S TESTICLE

How did males and females evolve? What was there before? Why do most species have only two sexes? When is it wise to be male and female both at once? When should you change sex? And what makes a boy a boy and, of course, a girl a girl? It's gender studies—but not as you know it.

*Dear Dr. Tatiana,*

*I'm a slime mold—*Physarum polycephalum *is the name—and I don't see how I'm ever going to marry and have children. I can only ooze along, so finding partners is difficult. I haven't met one yet. Worse, whereas every other species I've ever heard of has two sexes, my species has thirteen, and I gather that before you can make babies you have to convene them all. I don't see how this is possible, and I'm worried I'm going to end up a dreary old mold. Why are slime molds so oversexed?*

*Looking for a Baker's Dozen in the Forests of Romania*

Thirteen sexes? My poor dear, you are seriously deluded: your species has more than five hundred! But don't panic—you don't need to convene them all. Frankly, I think you need a quick lesson in slime mold facts of life.

First things first: what exactly is a slime mold? Most people, if they stumbled across you on a rotting log, would think you were a type of streaky yellow fungus. But nothing could be farther from the truth. Slime molds are a world unto themselves, only distantly related to animals, vegetables, or fungi. There are a couple of different types; you happen to be what's known as a true slime mold. A mature true slime mold is one enormous cell—easily visible to the naked human eye. However, whereas a regular cell is made up of one nucleus surrounded by a dollop of cytoplasm, the cell of a mature slime mold is a big mass of cytoplasm filled with millions of nuclei. This entity creeps along, devouring any microbes in its path. It does not copulate—and hence has no need to find a partner. Instead, when the time comes to have sex, it grows into a stalked fruiting body that looks like a tiny blistered lollipop. This structure releases spores into the air just as a flower might release pollen, and the spores develop into sex cells.

What's a sex cell? There's no mystery about that. A sex cell carries one complete set of your genes. Its mission is threefold: to find, recognize, and fuse with a suitable sex cell from the same species. What do I mean by "suitable"? I mean a sex cell of a different sex.

You see, sexes simply define who can reproduce with whom. Or, to be more precise, sexes define which sex cells can fuse with each other. Among animals, then, there are two types of sex cells, big (eggs) and small (sperm). A female or a male makes only one type; a hermaphrodite makes both. It's a truism that two sperm cannot fuse to form an embryo, and neither can two eggs. The only possible combination is one egg and one sperm.

Among slime molds, however, the situation is different. Along with organisms such as green algae, seaweeds, diatoms—tiny golden algae that secrete beautiful, symmetrical shells—and many of nature's more obscure creations, slime molds make sex cells that are only one size, a condition known as isogamy. When size is irrelevant, other features of a sex cell determine its sex. In principle, these could be anything—the simple presence or absence of a particular chemical on the surface of a cell would be one possibility. In your case, alas, it's nothing that straightforward.

Slime mold sexes are determined by three genes, known as *matA*, *matB*, and *matC*. Each of these genes comes in several variants—*matA* and *matB* each have thirteen known variants (which is presumably where you got the erroneous idea you have thirteen sexes), while *matC* has three. Now, bear in mind: as a mature slime mold, you have two copies of every gene. So let's suppose you have variants 1 and 3 of *matA*, variants 2 and 4 of *matB*, and variants 1 and 3 of *matC*. When you turn into the blistered lollipop and make your sex cells, each of them receives one complete set of genes—including one each of *matA*, *matB*, and *matC*. Each sex cell could thus receive any combination of the variants of these genes: one sex cell might have *matA1*, *matB2*, *matC1*; another might have *matA3*, *matB2*, *matC1*—you get the idea. As an individual slime mold, you are therefore capable of producing sex cells of eight different types—all combinations of your *A*s, *B*s, and *C*s. If you want to impress people, just call yourself an octosex. Out there in the forest, of course, other slime molds have different combinations of the variants of these genes; when you count up all possible combinations of *matA1–13*, *matB1–13*, and *matC1–3*, you get more than five hundred. (And indeed, since more variants of these genes probably await discovery, the slime mold sexes tally may go even higher.)

When your sex cells go forth into the world—and slime mold

sex cells are unusually independent; they are even able to eat (imagine a sperm stopping to have a snack)—their mission, as I've said, is to find a suitable partner to fuse with. And in your case suitable means that your partner has different variants of each of the three genes. So a *matA1*, *matB2*, *matC1* sex cell could fuse with a sex cell carrying *matA12*, *matB13*, *matC3*, but not with one carrying *matA12*, *matB2*, *matC3*. Complicated, eh?

All the same, your system is not freakish. On the contrary. Theories predict that isogamous organisms should have an embarrassment of sexes and that the really outlandish thing is to have only two. Let me walk you through the reasons why. Imagine the world from a sex cell's point of view. And imagine that you're in a population that has zero sexes—every sex cell can fuse with every other. In such a population, finding a mate is as easy as it gets. There is a drawback, however. Zero sexes cannot prevent inbreeding. If you encounter a sex cell from the same parent, you can go ahead and fuse with it.

Now imagine that you're in a population that has a large number of sexes. The larger the number of sexes, the easier it is to find a suitable partner to fuse with. At the same time, you become less likely to commit incest: sex cells from the same parent are much less likely to be allowed to fuse with each other. (In the slime mold case, a sex cell of a given type can fuse with only one-eighth of the other sex cells produced by the same parent.) In other words, a large number of sexes simultaneously increases the probability of finding a mate while reducing the risk of inbreeding.

How do you get from zero sexes to hundreds? Taking the first step—from zero to two—may be quite difficult, and exactly how it can happen is controversial. (Obviously the transition cannot be all that difficult, though, since we know it has happened repeatedly.) Be that as it may, once this initial hurdle is leapt, evolving more than two sexes is a snap. The reason is that if an individual starts making sex cells of a third sex, these will be able to fuse with sex cells of both the other sexes (but of course not with itself). At first, this new sex has an advantage because it can fuse with a larger proportion of the population than the other sexes can. The genes for the new sex will spread—until the population reaches an equilibrium where all three sexes are present in equal frequencies. If a fourth sex should then appear, the same thing would happen. Because being a new sex is always advantageous, the number of sexes will gradually drift upward.

And yet, as you correctly observed, most other isogamous organisms do have only two sexes—the least convenient situation with respect to finding a mate. Or, to put it another way, evolving more than two sexes is easy and has an obvious advantage; yet most isogamous organisms are stuck at two (though not male and female). Which suggests that there is some other power at work, something that strongly constrains the number of sexes that isogamous organisms can have. But what could it be? And how come slime molds are exempt?

No one knows for sure what the constraining factor is. The most likely candidate, however, is the need to control the inheritance of unruly genetic elements in the cytoplasm. You see, in addition to the regular genes of the nucleus, most organisms harbor other genetic elements: mitochondria, for example, or chloroplasts. (Chloroplasts are found within the cells of plants and green algae and are responsible for manufacturing energy from sunbeams; mitochondria are found within almost all cells, except those of bacteria, and metabolize carbon compounds.) These elements reside in the cytoplasm of the cell, often in extremely large numbers. They are thought to be the remnants of bacteria that once, long ago, were independent organisms. At some point in the ancient past, the bacteria took up with primitive cells, trading energy for shelter, and as time went by, they became unable to function on their own. Now, they have just a few genes left to them, a rump genome. But, as everyone knows, rumps can be troublesome.

Problems are most likely to arise if mitochondria and chloroplasts are inherited from both parents. Then, the mitochondria (say) of the parents may compete with each other in ways that harm the organism. For example, mitochondria from one parent could attempt to exclude mitochondria from the other parent from the sex cells; such goings-on may result in the mitochondria being less efficient at what they're meant to be doing—metabolism. The easiest way to prevent any problems of this sort is to make sure that mitochondria—and chloroplasts, if you've got them—are inherited from only one parent.

Why would this constrain the number of sexes to two? Well, the idea is that strict control over the inheritance of these elements is overwhelmingly important, and by far the simplest method of control is to have one sex that never gives them and one sex that always does. A number of isogamous organisms have mechanisms to ensure that only one parent will pass on these elements. The green alga *Chlamydomonas reinhardtii*, for example, has sex cells of two types, known as plus and minus. The plus type transmits chloroplasts, the minus type mitochondria.

There's another piece of circumstantial evidence that controlling these elements constrains the number of sexes. Two groups of organisms—mushrooms and single-celled critters called ciliates—have sex not by making sex cells but by two individuals swapping half a nucleus. (This system has the odd effect of making you become genetically identical to a total stranger—the two of you are suddenly identical twins.) Crucially, no cytoplasm changes hands, and these organisms have no need to regulate the inheritance of mitochondria. Sure enough, in these groups the number of sexes can be opulent. *Schizophyllum commune*, a pinkish, hairy mushroom that grows on tree trunks, has as many as twenty thousand different sexes.

So you see, what really makes slime molds exceptional is not so much the number of sexes but the fact that you have lots of sexes *and* you merge the cytoplasm from two parents. How do you get away with it? Are your mitochondria better behaved than those of other species? Nope. The key is that your mitochondria are still inherited from only one parent. The gene *matA* supervises the elimination of mitochondria from one parent or the other. There's a hierarchy of the variants so that if cells carrying *matA12* and *matA2* fuse, the mitochondria that came with *matA12* will be destroyed, whereas if *matA12* and *matA1* fuse, the mitochondria accompanying *matA1* will get the ax. I imagine that this system doesn't evolve often because it is difficult to get it to work—and I salute slime molds everywhere for making it a success!

౷

*Dear Dr. Tatiana,*

*In the whole of my species, the green alga* Chlamydomonas moewusii, *I think I'm the only male: I make small, elegant sex cells, everyone else makes big, clumsy ones. But being the only male is no bed of roses. The big, clumsy sex cells can fuse with each other, so I'm not especially in demand; in fact, I suspect that my sex cells are being discriminated against. It's just not fair. What's going on?*

*Ready to Litigate in Tallahassee*

I suspect you're just a man whose time hasn't come. Rather than being the first father of a great tribe of green algae, you'll probably die without begetting anybody. Why? Well, your species is isogamous—and you are just a mutant who produces smaller sex cells than everyone else. When a sex cell of yours fuses with a regular sex cell, the resulting cell, known as a zygote, will be smaller than usual. This will almost certainly reduce its chances of survival.

You've been unlucky. Species containing males and females have evolved from isogamous species over and over again. Indeed, although isogamy has long gone out of fashion for plants and animals, the distant ancestors of both groups are thought to have been isogamous. So what's the secret of making the transition from being an isogamous species with two sexes to having males and females? Let me say at once that there are lots of ideas—but no definitive answers.

The essential problem in the evolution of males and females is to imagine forces that favor individuals who make either big sex cells or small sex cells but not sex cells of intermediate size. At first glance, it's easy to imagine how an individual could benefit from making small sex cells—small sex cells can be produced in large quantities, and an individual who makes more sex cells than his rivals increases the number of other sex cells he can potentially fuse with. The problem is, though, as you've found, producing lots of small sex cells won't get you anywhere if these small cells ruin the zygote's chances of survival. Indeed, if we make the reasonable assumption that for a zygote to be viable it must be at least as large as the zygote that results from the fusion of two isogamous sex cells, then males and females can only evolve if, at the same time that some individuals are starting to make smaller cells, other individuals are starting to make bigger ones.

But what force can favor making fewer, larger sex cells? This is much harder to envision. The best guess is that in the splash and dunk of the ocean—where the first males and females must have swum—larger cells are easier to find, either because they are easier to bump into or because they are more efficient at dispersing chemical attractants—better at shouting "I'm over here, I'm over here."

If you've read my correspondence with the slime mold, however, you may be wondering whether the evolution of small or large cells could have anything to do with controlling the inheritance of mitochondria or chloroplasts. It's tempting to think so. After all, eggs come with lots of cytoplasm; sperm don't. So perhaps she passes boisterous genetic elements and he passes the buck.

Then again, perhaps not. It's true that in most species that have males and females these elements are inherited from only one parent. That parent, however, isn't necessarily the mother. And there's no evidence that controlling these elements has anything to do with the evolution of eggs and sperm. No, my man, I fear mitochondria are the least of your worries. Get yourself a sex-discrimination lawyer if you want, but don't expect to win.

༄

*Dear Dr. Tatiana,*

*We're sea hares of the species* Aplysia californica. *We've been having a fabulous orgy—being both male and female, we all get to play both roles at once. That's right, each of us plays male to the sea hare in front and female to the one behind, and often the party rocks on for days! It's such a great system, so much better than being male or female, that we're mystified why*

*everyone hasn't followed our lead. Why aren't all living things*
*hermaphrodites?*

*Group Sexists in Santa Catalina*

Orgies beneath the waves! I can just see it: a cozy, copulating
chain of beautiful creatures that resemble snails who've lost their
shells and are hiding the fact with delicate colored folds of membrane. You sea hares are obviously so intent on your orgies that
you're ignorant of the ways of other hermaphrodites. Not all
hermaphrodites have orgies—they get up to all kinds of other
hanky-panky instead. Let me give you a sampler.

Exhibit A: the black hamlet fish, a small carnivorous fish that
lives in the tropics. An hour or two before sunset, any fish in a
sensual mood will cruise the edge of a reef and find a mate. The
couple will then take turns playing the male or the female, swapping roles after each bout of spawning.

Exhibit B: *Diplozoon gracile*, a parasitic fluke that lives on the
gills of fish. Mating pairs turn into a fluke version of Siamese
twins: they actually fuse together permanently, their genitalia
in contact in perpetuity. I hope they never take a dislike to each
other.

Exhibit C: the European giant garden slug, *Limax maximus*. To
prepare for a tryst, two slugs sit together on a tree branch and
secrete mucus for about an hour—pretty kinky. Then they wrap
their bodies together and dive headfirst off the tree. Instead of
plunging to the ground below, they hang in the air, suspended on a
mucus rope. As they dangle upside down, they unfurl their pale,
ribbonlike penises from the sides of their heads. The penises also
dangle (about three centimeters below the loving couple), en-
twined, their tips pressed together to exchange sperm. Neither
partner actually penetrates the other—and the session lasts for

hours. In *Limax redii*, a close cousin, which likewise has its penis on its head and conducts sex upside down in midair, the penises dangle a full eighty-five centimeters, the length of three champagne bottles end to end. Talk about a head rush.

Nonetheless, you ask an intriguing question: why aren't we all hermaphrodites? Or, to put it another way, when is it better to be a hermaphrodite than to be a girl or a boy? The answer has nothing to do with orgies, I'm afraid.

In the most general terms, hermaphrodites are predicted to evolve whenever the payoff—measured in nature's usual currency, children—is greater for hermaphrodites than for individuals of only one sex. And when is this the case? Well, hermaphrodites may have an advantage if individuals live at low density, a situation that can make it difficult to find a mate. A hermaphrodite might be able to self-fertilize, and even if it can't (or doesn't want to), it will in principle be able to mate with everyone it meets. Alternatively, hermaphrodites may be better off when the extra trouble of being both sexes is negligible. Plants pollinated by bees or other creatures have to make beautiful, showy flowers whichever sex they are—so they incur little additional cost if they are both. Indeed, if pollinators are primarily attracted by pollen, then flowers that are exclusively female—and thus produce no pollen—may have no visitors. In contrast, plants pollinated by wind are limited not by how many pollinators they attract but by how much pollen they spread about or how much fruit they produce. In this case, then, individuals may do better if they specialize, making either pollen or fruit.

How closely does the theory match the data? Well, flowering plants pollinated by wind do tend to be one sex or the other whereas those pollinated by animals do tend to be both. Beyond that observation, however, picking out trends is extremely difficult. Hermaphrodites are even more diverse than their sexual practices. They are found among most major groups of animals, from flatworms to fish, and among many groups hermaphroditism is the norm. In some groups, however, hermaphrodites are rare or absent. Of all the thousands and thousands of insect species, there are just one or two known hermaphrodites, for example. And among mammals, birds, reptiles, and amphibians, hermaphrodites are unheard of. More perplexing still, when you look at the distribution of hermaphroditic animals, there are no clear ecological associations (such as low density) that reliably occur with the trait. Among fish, for example, most species have males and females. Nonetheless, where hermaphroditism has evolved, it has done so in species that live in wildly different environments. You may recall our friend *Rivulus marmoratus*, who makes its home in mangrove swamps. Then there's the aforementioned black hamlet fish, along with its relations, all tropical reef fish. And then there are the deep-sea hermaphrodites. In contrast, all comb jellies are hermaphrodites—*except* for a few species that live in the deepest seas. Or worse, compare bivalves—animals that have hinged shells, such as clams and mussels—with barnacles. Among bivalves, parasitic species tend to be hermaphrodites, while respectable free-living types tend to be one sex or the other. Among barnacles, however, it's the reverse: free-living species are typically hermaphrodites, parasites have separate sexes. What's going on?

There are three reasons the picture is so confused. First, evolving from separate sexes to hermaphroditism—or vice versa—may not be easy. This is not something we know much about. But certainly to evolve from being one sex to being a hermaphrodite, an organism has to evolve a second functioning reproductive tract. Achieving this may be tricky, requiring several unlikely genetic events. Making such a transition is surely easier in some groups than in others. And even if the transition is genetically easy, other considerations may stand in the way. For example, hermaphrodites may have behaviors that stop separate sexes from catching on. In hermaphrodite societies—such as that of the black hamlet fish—where mating is based on the reciprocal trading of eggs or sperm, an individual who is only one sex may be discriminated against as a partner because he or she cannot play the game.

The second reason for so confusing a picture is that such a variety of circumstances favors hermaphrodites: there's no trademark hermaphrodite lifestyle. Finally, just where you might expect the emergence of hermaphrodites—as in low density—other, equally successful solutions may emerge. For example, being a hermaphrodite might not be the only way to find a mate when mates are few and far between. Instead, all candidates could assemble at a fixed time and place. The extraordinary mass spawnings of many marine animals, where millions mysteriously know to convene in the lagoon when the July moon is full (or whatever the signal happens to be), might be an instance of this. And I'd guess the dearth of hermaphroditic insects is due to the evolution of equally fruitful alternatives.

I'd like to leave you with this last thought. Not all species have either males and females or hermaphrodites. Some have hermaphrodites and males. Some have hermaphrodites and females. And a few daring sorts—such as the Mexican bat-pollinated cactus *Pachycereus pringlei*—have some of everybody. Now that gives a tingle to "ménage à trois."

∽

*Dear Dr. Tatiana,*

*There's been a frightful accident. I was happily sitting in my usual spot at the bottom of the sea when I felt an itch on my nose. Being a green spoon worm, I don't have arms and I couldn't*

*scratch. So I sniffed. And I inhaled my husband. I've tried sneez-*
*ing, but he hasn't reappeared. Is there anything I can do to get*
*him back?*

*Too Much Heavy Breathing near Malta*

There, there, it's no use crying over snuffled husbands. He
wanted to be snuffled, and he's not coming back. By now he'll
have assumed his position in your androecium—literally, "small
man room"—a special chamber in your reproductive tract where
he can sit and fertilize passing eggs. How does he fit? The little
chap is 200,000 times smaller than you: it's as if a human male
were no bigger than the eraser on the end of a pencil. You could
keep a score of husbands without trouble.

But you mustn't disdain your diminutive lover. It's only by
chance that you escaped his fate. You see, when a green spoon
worm larva first hatches, it has no sex. Instead, its sex is determined by the events of its first days. If, during this time, the larva
encounters a female, it becomes male. If, after about three weeks,
it hasn't met a female, it settles into a comfortable crevice and
becomes female itself.

This probably sounds amazing—and in many ways it is. However, before talking about the strangeness of your sex life in more
detail, I'd like to draw your attention to a phenomenon that's
even more peculiar. You'll probably agree that "male" or "female"
is one of the most basic attributes an organism can have: after all,
males and females reliably occur in millions of species. So you
might imagine that the way a creature becomes male or female
varies little from one species to the next—and that your situation
is unique. You'd be wrong on both counts. Surprisingly, an organism's sex is determined in ways that vary enormously. And you

green spoon worms aren't the only ones whose sex is determined by social milieu.

Broadly speaking, sex is determined either by genetic or by environmental factors; within these two categories, however, there are all sorts of possible variations, many of which have evolved over and over again. For example, one of the most common ways sex is decided is by special chromosomes. Among mammals, males have an X and a Y chromosome, females have two X chromosomes. For birds, the situation is reversed. Males have two Z chromosomes, females have one Z and one W. Fruit flies have XY males; butterflies have ZW females. Lizards swing both ways: some species have ZW females, others have XY males. It's crazy. And that's just chromosomes. I haven't even mentioned all the critters where the males hatch from unfertilized eggs—a system thought to have evolved at least seventeen times—let alone species where sex is determined through horribly complicated interactions of many different genes.

What about environmental factors? For many reptiles, what matters is the temperature at which their eggs are incubated. Thus, in alligators (and in still more lizards), you get girls when eggs are buried in cool sand and boys when the sand is warm; for many turtles, it's the other way around. Snapping turtles and crocodiles are even wackier: eggs buried either in mounds of cool sand (around 20°C or 68°F) or hot sand (40°C or 104°F) hatch out girls, eggs buried in warm sand hatch out boys. More curious still, in *Stictococcus sjoestedti*, a tropical insect that sucks the sap of cocoa trees, eggs infected with a particular symbiotic fungus become females, uninfected eggs become males. And then, there are those like you whose sex is determined by social circumstance.

For many individuals, this involves changing sex. In one species of *Capitella*, a worm partial to sewer sludge, males turn into hermaphrodites if they fail to encounter a female within a certain time. In the slipper limpet *Crepidula fornicata* (a notorious pest of oyster beds), everyone starts his career as a male. A fellow who finds himself alone, however, quickly turns into a female and starts attracting mates. Other slipper limpets pile on, gradually forming a louche limpet stack. In slipper limpet sex, it's males on top: although small, they have splendidly long penises so they can fornicate with the female at the bottom. But as the stack continues to grow, the guys who were once at the top of the heap find themselves in the middle and change sex, resorbing their penises in the process, to become female. More exotic: the marine worm *Ophryotrocha puerilis*. If two females find themselves together, the smaller one changes into a male. But because females grow at a slower pace than males, the male will soon become the larger member of the pair. At this point—shazam!—both individuals change sex. Such reversals happen repeatedly. In the end, though, pairs that have been together for a long time end up by both turning into hermaphrodites. An enviable life.

As a general rule, flexible gender is expected to evolve whenever an individual's reproductive success as a male, female, or hermaphrodite differs greatly according to circumstances. Social milieu may not be the sole influence at work—if, for example, males can't reproduce successfully unless they are big, it could be advantageous to start life as a female and become male only on achieving a good size. The ability to choose your sex is particularly handy, however, when being one sex leads to a riskier life than being the other.

Which brings me back to you. A female green spoon worm takes a greater gamble in life than a male does. She needs two years to mature, during which she may be eaten by a bat ray, and on reaching adulthood she may never find a mate. So it makes sense for a larva who meets a female to become male: not only is he guaranteed a mate, but he can start reproducing as soon as he's installed himself. What is it about you that makes a larva want to be a man? Well, your lovely bulbous body—but particularly your long, twitchy proboscis—secretes a substance known as bonellin, after your formal name, *Bonellia viridis.* A whiff of bonellin makes any larva stand up and fly right.

But what you're really dying to know, I suspect, is why your lovers are so minute: what strange circumstances prompt natural selection to reduce a man to a testicle? Two factors are thought to be conspiring here. The first is if females are sedentary, the second if they are sparsely sprinkled across the landscape. Then a male's biggest challenge is finding a mate. The smaller he is, the faster he can mature (he doesn't have to waste time growing) and the sooner he can start looking.

This size business is not just a quirk of green spoon worms: lilliputian lovers appear in widely separate groups. Take anglerfish, monsters that live in the coldest, deepest seas. The females don't swim much but float in the darkness, ready to ambush their prey. Like wreckers of old, these formidable girls have special dangles and lanterns to lure the curious to their doom. Victims are swallowed whole, engulfed by toothy mouths and grossly distendable stomachs. Like your hubby, male anglerfish are minute. But these guys win the all-species Cyrano de Bergerac Award for the largest nose in proportion to body size. Presumably, the males follow their noses to find females in the vasty deep. When they meet one, they bite into her leathery black underbelly and fuse with her body, becoming a permanent appendage, little more than a pair of gonads. Still, it seems to me, their fate isn't quite as ignominious as a life sentence in your small man room.

∽

*Dear Dr. Tatiana,*

*I'm a spotted hyena, a girl. The only trouble is, I've got a large phallus. I can't help feeling that this is unladylike. What's wrong with me? Can anything be done?*

*Don't Wanna Be Butch in Botswana*

No one expects hyenas to be ladylike. Least of all the biggest and baddest of them all—the spotted hyena. The brown hyena and the striped hyena squabble with vultures over a rotting carcass, and both sometimes eat fruit. The aardwolf, a dainty black and white hyena, eats harvester termites, lapping up more than 200,000 a night with its sticky tongue. But the spotted hyena is a fearsome predator. A single hyena can run down and kill an adult male wildebeest, an animal more than three times its weight. And despite their respective reputations, lions scavenge from hyena kills more often than the other way around. At least, they do if they get there in time. A spotted hyena can devour a Thomson's gazelle fawn (2.5 kilos or 5.5 pounds) in under two minutes. Twenty-one hyenas can dispose of a yearling wildebeest (100 kilos or 220 pounds) in thirteen minutes, and there won't be much trace of him left. Having massive jaws, hyenas can pulverize bones, even rhinoceros bones, and not just to get to the marrow: unlike other carnivores, hyenas can digest bones. That's why hyena scats are white—they're mostly bone powder. All the same, the first hyena on a kill will begin by taking delicacies such as the victim's testicles or udders, or a fetus if there is one.

So you see, yours is no tea-drinking, cake-eating, genteel society; ladies would be distinctly out of place. And not just with respect to table manners. Spotted hyenas typically live in big groups, each presided over by a dominant female. But while they

sometimes hunt in packs (especially when hunting zebra), cooperation on anything else is rare. Instead, it's a mad scramble.
That's one reason hyenas eat so fast—they gobble as much as
they can before anyone else arrives. The only moderating influence is the social hierarchy, aggressively enforced, where the
dominant female and her cubs take precedence over everybody—
and all other females take precedence over all other males. Unlike
the other hyena species, where males and females are roughly the
same size, but like many birds of prey, female spotted hyenas are
bigger and heavier than males. But a phallus? At least no one can
accuse you of penis envy.

On the outside, male and female genitalia look so similar that
for many years the spotted hyena was thought to be a hermaphrodite. In females, however, what looks like a phallus is actually a
grossly enlarged clitoris, fully capable of erection. The lips of the
vagina have fused shut and form a pseudo-scrotum. Urination,
copulation, and birth must, therefore, be done through the
clitoris.

How? Well, if you really want to know . . . At puberty, the
mouth of the clitoris becomes elastic, able to open to about two
centimeters (one inch) in diameter. To allow copulation, the female
retracts the clitoris, folding it up like a concertina, thus creating
an orifice and allowing the male to slide in. It's the birth of a spotted hyena, however, that is particularly bizarre. For starters, the
birth canal is a funny shape. Instead of being a straight passage as
in most mammals, it features a sharp bend. Worse, at sixty centimeters (twenty-three inches), it's twice as long as in other mammals of a similar size. The umbilical cord, however, is short—only
eighteen centimeters (seven inches) long. Once the placenta de-
taches, the young hyena will asphyxiate if it is not born promptly.
But a baby hyena's head is too big to pass through the clitoris. So
when a mother gives birth for the first time, the clitoris tears to

let the cub out. This is not just agonizing. It is often lethal. Scientists estimate that more than 10 percent of females die the first time they give birth, and more than half of firstborn cubs are stillborn. (Paradoxically, since the clitoris never recovers from this trauma, subsequent births don't put the mother's life at risk.)

So here we have a strange set of facts. Chief among them is that the female spotted hyena has a reproductive organ that exacts huge costs. This demands explanation. Either the organ itself must confer some giant benefit or it must have evolved as an unfortunate by-product of some *other* trait that confers a giant benefit.

Let's start with the organ itself. Two advantages have been proposed. The first is that in having a structure that mimics the phallus, the females can take part in greeting ceremonies: when spotted hyenas meet, they stand head to tail and inspect each other's erect members. Female participation in the ritual might therefore help them exert their dominance over males. Although this idea will appeal to any Freudians out there, it hardly seems sufficient to explain such a deadly organ.

The second possibility is even more flimsy. The mechanics of spotted hyena sex are so tricky that females are able to resist unwanted advances: copulation requires full cooperation; rape is impossible. But rape has never been reported in any hyena species. Moreover, since female spotted hyenas are so much bigger than males and have such big teeth, courting males are unusually polite, literally bowing and scraping as they approach. Females hardly need a phallus for self-defense.

Neither of these explanations is exactly convincing, I think you'll agree. What about the phallus being a by-product of natural selection for Something Else? At first glance, this seems more plausible. There's a good candidate for the Something Else: aggression. We know hyena society is aggressive, and it's easy to imagine that aggressive females do better than shrinking violets. Moreover, the fetal hyena is exposed to high levels of testosterone and other androgens—the "masculine" hormones—while in the womb. Intrauterine exposure to these hormones fosters aggression: female mice that are snuggled between their brothers during fetal life are exposed to higher levels of androgens than females nestled between sisters and are more aggressive as grown-ups. And crucially, intrauterine exposure to high levels of androgens can cause profound genital abnormalities. In humans, for example, excessive exposure to androgens in the womb gives a girl a greatly enlarged clitoris and a vagina that is partially fused shut. So the question is, could increased aggression in female spotted hyenas be sufficiently favored by natural selection to offset the costs of copulating and giving birth through the clitoris?

Maybe. Spotted hyenas begin as they mean to go on: with their teeth bared. Most hyenas are born in litters of two, and whichever is born first will attack the second within minutes. Death often results. Killing your sibling allows you to monopolize your mother's milk; since spotted hyenas nurse for more than a year, successful siblicide increases your own chances of living to adulthood. Thus, one idea is that high levels of androgens in the womb are favored because they promote violence among cubs at birth. This fails to explain, however, why siblicide is more frequent among same-sex pairs than among pairs of opposite sex and why it is even more frequent among pairs of females than among pairs of males: if siblicide were the explanation, you wouldn't want any sibling, whatever its sex.

A more convincing explanation says that aggression is favored because dominance relationships are mediated by aggression—and there are considerable benefits to being dominant. Compared with their more lowly peers, high-ranking females become pregnant younger and have shorter gaps between litters, and more than twice as many of their offspring survive to adulthood. This is a big difference and could potentially offset the cost of the phallic clitoris.

Alas, however, the puzzle can't be solved so neatly. Studies of the spotted hyena clitoris show that blocking the circulation of androgens in the womb does not cause reversion to "typical" female genitalia. A large component of the development of the phallic clitoris is thus independent of these hormones, undermining the idea that the phallus is a by-product of natural selection for increased aggression. So until we know more about how the structure develops, I'm afraid the reason for your singularly costly phallus will remain a mystery.

But your situation does illustrate a more general point. That is, beyond the basic fact that males make sperm and females make eggs, there are no rules, not even in what appear to be the most stereotypical gender-related areas. Let me give you two examples—genitalia and child care.

In countless groups of animals, females have evolved internal fertilization, presumably because it is an effective way of ensuring that egg meets sperm. Internal fertilization can be achieved by the female's squatting over a sperm packet, as it is in some mites and some amphibians, for example. But often the evolution of internal fertilization is accompanied by the evolution of a penis, a structure to deliver sperm. The penis has been reinvented more often than the wheel. Which explains why, in different groups, this symbol of masculinity is formed from different parts of the body—heads, mouths, legs, tentacles, fins, and so on. Some of the reinventions are pretty quirky. The spider, for example, is stuck with the penis equivalent of a triangular wheel. The male, as you may recall, delivers sperm with pedipalps, modified mouthparts. Inconveniently, however, the pedipalps have no connection to the part of the body where sperm is made, so before copulating, the male deposits a drop of sperm on a small web that he spins for the purpose. He then draws the sperm up into his pedipalps, like someone drawing ink into a fountain pen. In the seahorse, it's the female who has the penis, to deliver eggs into the male's brood pouch. One species of sea slug, *Sapha amicorum*, a tiny hermaphrodite from the Red Sea, actually has its male genitalia inside its mouth, and copulation is an extraspecial kiss. Lucky they don't have to go to the dentist. But perhaps the oddest approach I've come across is practiced by three obscure relations of the octopus, all of which have abandoned the seafloor for open water. The paper nautilus, the best known of the trio, is an ethereal creature. Bright white, with hints of purple, blue, and red, the female lives in a beautiful white shell and floats through the water. The male is tiny, and hardly anyone has ever seen him. Not even his mate. What appears to happen is that he fires off his penis—a modified tentacle—which takes up an independent life within the female, who may entertain several such guests at once. This is so weird that it's not surprising early naturalists thought the penises were parasitic worms. Imagine the lonely-hearts advertisement of a female paper nautilus: "Fire and forget. Send your organ to a loving home."

Child care is another of Mother Nature's favorite inventions. It has evolved, to different degrees, in an astonishing diversity of organisms, and mothers have no monopoly on the activity: depending on the species, the carer may be hermaphrodite, male, or female. Take leeches. These bloodsucking hermaphrodites often perform rudimentary parental care, guarding their egg cocoons from predators. But some go further. The African leech, *Marsupiobdella africana*, has adopted the habits of a kangaroo and carries its young in a pouch. And the leech *Helobdella striata* not only carries its young glued to the belly of the parent but hunts small worms for them to eat. Or take frogs. Most frogs spawn, and that's it. But in a few, child care is elaborate. Green poison arrow frogs go to great lengths for their tadpoles. These small, elegant creatures live in the leaf litter of Central American forests. As their name suggests, their chief claim to fame is that their skins are toxic. Humans living in the forests wipe their darts on the frogs' backs, collecting venom to paralyze the animals they hunt. However, these frogs deserve fame for another reason: they are a paragon of fatherhood. Once a male and female have courted, the female lays a small clutch of eggs in the leaf litter. The male tends the eggs, sitting in a puddle, then returning to sit on the eggs to keep them moist. He uses his puddle jaunts to search for pools where he can drop the tadpoles when they are ready. A pool might be the accumulation of rain in the top of a pineapple or a cranny of a tree trunk. When, after a couple of weeks, the tadpoles hatch, he carries them, one at a time, to his chosen pools and drops them in.

The variations on these themes are myriad, fascinating, shifting like shards of colored glass in a grand kaleidoscope, often defying prediction. My favorite example is the spraying characid. This little fish lives in murky rivers in Guyana. Surprisingly for a fish, it lays its eggs out of water. When the male and female spawn, they leap out of the water together and stick themselves briefly to a blade of grass or to the underside of a leaf from a plant hanging over the bank. With each leap, the female lays eggs and the male fertilizes them. They do this over and over again until the female has laid perhaps three hundred eggs. Then, for the next three days, the male splashes the eggs with his tail to keep them from drying out. If it's raining, he gives himself the afternoon off. Or, to take an example more similar to your own case, my hyenid friend, look at the Dayak fruit bat, who lives on the Malay Peninsula. In this species, both males and females produce milk, apparently sharing the responsibility of nursing the young. Is it any stranger for a female to have a phallus than for a fish to lay eggs out of water? Or for a male mammal to produce milk?

∽

**Gender benders.** The next time anyone wheels out a stereotype and says, "She does this, he does that," here's your reply:

When you gaze at a couple and wonder
What trait makes him "him" and her "her,"
Beware, for it's easy to blunder
And be false in what you aver.

Some creatures change sex before teatime,
Some others find two sexes dull,
And that virile male fish has no free time—
He's got all his kiddies to lull.

When it comes to the topic of gender,
Mother Nature's been having some fun.
Take nothing for granted! Remember,
You won't find any rules—not a one!

# WHOLLY VIRGIN

No doubt, many of you have been writing to me over the years because you've seen my popular TV program, *Under the Microscope—The Deviant Lifestyle Show!* You'll know that the program's had a lot of kinky, if not downright perverted, guests, and of course the audience is used to hearing about weird sexual practices. A few weeks ago, however, we had a really provocative guest. I don't know if you saw the episode, but the most awful fight erupted—there was practically a riot, and I'm sorry to say I almost lost control of the show.

I'm not surprised everyone got so upset. The guest in question doesn't have any weird sexual practices—or indeed, any sex at all. Worse, no one in her family has had sex for more than eighty-five million years. This is an outrage: scientists cannot agree on what sex is for, but they all agree that it is essential, impossible to live without. And yet if our guest can manage without sex—or men—why can't the rest of us? What's sex good for? Is it passé? Are men endangered? Such questions bring us to the heart of perhaps the most fundamental, controversial matter in biology—what is sex for?—and since most aspects of the topic were hashed out on the show, here's my account of the furor.

We had the biggest turnout yet. All the usual crowd were there—the belligerent ram and his supercilious armadillo friend were both sitting in their regular places in the front row. The pocket mouse was curled up in the small-animal gallery; the homing pigeons were on their customary perch. As always, Moby, that impressive puffer fish from the Congo River, was churning up and down the freshwater fish tank. But a lot of the members of the audience were first-timers. I noticed a bake—if that's the right collective noun—of clams in a corner of the saltwater tank. Several Brazilian lizards, looking rather sickly and depressed, had stationed themselves halfway up the left wall. The back rows were bristling with radical feminists wearing T-shirts that said, "Men, who needs them?" and "Sex is for wimps." The studio was fizzing with excitement and hostility. The cause of all the agitation? My guest, none other than Miss *Philodina roseola*, the bdelloid rotifer.

From looking at her, you'd never guess Miss *Philodina* is at the heart of one of the most notorious scandals in evolution. Slender and translucent, she looks less like an animal than like a pocket telescope blown out of pale pink Venetian glass. But you don't normally see telescopes eating algae, and before the show Miss *Philodina* had clearly been doing just that. Throughout the entire evening, the remains of a recent feast were greenly visible through her glassy body. (This was my fault: I usually advise translucent guests to skip lunch the day of the show, but this time I forgot.) Her most striking feature, though, is on top of her head, where she has a pair of disks edged with beating cilia, tiny mechanical hairs whose motion gives the illusion of wheels spinning round and round. Of course, she's barely half a millimeter tall, so before anyone could get a look at her, we had to settle her on a comfortable frond of moss and turn on the microscope so as to project her blown-up image on the screen by my chair. We cut a fine pair, the two of us. I looked my usual glam self in my best scarlet suit. She looked the picture of innocence. And that innocence was the cause of all the trouble . . .

The show started off much as usual. I welcomed the audience and introduced Miss *Philodina*, giving a few trivial details, such as her favorite spot to hang out (damp moss) and the fact that rotifer means "wheel bearer." Everyone in the audience applauded on learning that her Latin name means "rosy lover of twirling." But as soon as I began to talk about the nature of her deviancy, the disturbance began.

Me: Tell us, Miss *Philodina*, when was the last time anyone in your family had sex?

Miss P.: Hmmm. I think it's been about eighty-five million years since anyone in my family has even been on a date.

Me (To audience): And you thought you had problems. (To Miss *Philodina*): No sex, not even a kiss, since before the dinosaurs went extinct. And why not?

Miss P.: My ancestors abolished males. They said they were better off without them.

The studio drowned in jeers and whistles—despite loud cheering from the radical feminists.

Me: So how do you reproduce?

Miss P.: We clone ourselves.

Well, that caused absolute mayhem. The pocket mouse even fainted. But I've seen this reaction before. Many animals, especially mammals, have a horror of cloning. They seem to think it will produce a flock of monsters or something. So I had to remind everyone that cloning is nothing more than reproducing without sex—something that billions of respectable organisms are doing every day. I paraded my standard examples: strawberries sending out runners and shoots; yeasts and other organisms budding off bits of themselves; sponges, sea anemones, and worms of various kinds falling to pieces and regenerating, whole new animals growing from each piece; all sorts of girls (including bdelloid rotifers) laying asexual eggs. And to the discomfiture of many in the audience, I pointed out that even mammals clone once in a while, when an embryo splits early in development. The resulting individuals are not usually called clones, though. "Twins" is considered more polite. Typical mammalian political correctness.

It's curious. Everyone always forgets there's nothing wrong with cloning from time to time—that cloning mixed with bouts of sex can contribute to a healthy and happy lifestyle. As I explained to the audience, it's only giving up sex altogether that's the problem.

At first glance, however, giving up sex seems advantageous—from a genetic point of view, anyway. Sex may be fun, but cloning is much more efficient. All else being equal, an asexual female who appears in a population should have twice as many offspring as her sexual counterpart. To see why, think of it this way. In a sexual population—the human population, for example—each female must have two children for the population to remain the same size. If females have fewer than two children, the population shrinks; more than two, and the population grows. In an asexual population, however, each female needs to have only one child for the population to remain the same size. More than one, and the population will grow.

But although asexuality often evolves—it pops up in groups from jellyfish to dandelions, from lizards to lichens—it rarely persists for long. On the great tree of life, asexual groups are out on the twiggiest twigs of the twiggiest twigs: lots of buds, no branches. After a brief and glorious flowering, asexuals vanish. Which has led scientists to conclude that exclusive asexuality is an evolutionary dead end, a fast track to extinction. Sex, they insist, is essential. And ancient asexuals—creatures such as the bdelloid rotifers that have lived without sex for millions of years—should not exist. According to all the theories, the bdelloids should have disappeared shortly after giving up sex.

Yet, in scandalous defiance of scientific prediction, there's Miss *Philodina*, thumbing her wheels at us from the frond of moss. How come these rotifers have succeeded where so many others have failed? Or, to get back to the central question, if they can do without sex, why can't the rest of us?

With that preamble, I opened the floor to questions, as usual reminding the smallest members of the audience to step up to the microscope in the aisle. Understandably, the first questions challenged Miss *Philodina*'s claim. What did she mean when she said no one in her family has sex? Did she simply mean that bdelloid rotifers get intimate but avoid genital contact? No, that's not what she meant. But before she could go on, there was a truly embarrassing incident. Two bacteria tried to have sex on live TV.

The screen on the wall lit up: someone in the audience had gone to stand under the microscope. Gradually, the image came into focus to show not one but two lozenge-shaped organisms. In actual size, each would've been about a millionth of a meter—tiny even in comparison to Miss *Philodina*.

One of them started to squeak: "Good evening, everybody. We're a pair of bacteria of the species *Escherichia coli*—*E. coli* to our friends. Many scientists keep us as pets, so we often live lives of luxury and ease in the laboratory. In the wild, we live in the guts of mammals, helping digest food.

"For us bacteria, reproduction is reproduction and sex is sex. Unlike you 'higher' creatures, we're not so vulgar as to do both at once. For us bacteria, reproduction is asexual: we simply divide into two genetically identical cells. This way, sex—by which I mean the acquisition of extra genes—is something that we reap the benefits of during our lives. If humans could do this, which they can't, it would be like suddenly adding a few genes for longer legs or bluer eyes."

At this, one of the homing pigeons muttered wryly, "Lucky bacteria. Getting new genes beats a tummy tuck for a midlife crisis. Wish I could get some new genes."

Before I could turn the discussion back to Miss *Philodina*, the bacteria said, "So, how we do it? We have several ways. We pick up DNA that's loose in the environment. We gather DNA from passing viruses. We even plunder the genes of dead bacteria—the cognoscenti call it necrophilia."

The thought of necrophilia—even if only bacterial necrophilia—produced gasps of horror from the audience, and someone yelled, "Perverts!"

The bacterium blithely went on: "We also indulge in bestiality, getting genes from bacteria of other species. But above all, when we're in the mood, we have sex with each other. We'll show the rotifer how it's done—I'm going to give my friend here a useful set of antibiotic resistance genes. Look, Miss *Philodina*, watch how we do it!"

As the crowd roared, one of the bacteria started to extend a tube toward the other.

Luckily, at that moment my technician flicked the microscope switch and plunged the screen back into darkness. I tell you, it was a close call. Talking about sex is one thing; as you can imagine, showing sex would land me in all sorts of hot water with the network and thoroughly upset my commercial sponsors (though *PlayBeast* would've been thrilled). It could've been the end of the show.

The dirty little germs did do me one favor, though. They reminded everyone what sex is—and that it's not the same as reproduction. Sex is any process that mixes genes from different individuals.

To my amazement, Miss *Philodina* took up the cue. Like a pubescent boy, she knows an awful lot about the theory of sex; unlike a pubescent boy, she obviously finds the whole idea disgusting. I guess that's bound to happen if you've got eighty-five million years of celibacy behind you.

"Bacteria," she said, sighing. "Always exaggerating their sex lives."

She was right, of course. Despite the impression the speaker tried to give, bacteria are not nature's libertines.

"Most bacteria don't have sex of any sort very often," she went on, "and *E. coli* are among the most abstemious." She spun her wheels indignantly. "I wouldn't want anyone to think that we bdelloids are merely a sort of bacteria. Or worse, a sort of virus."

She was worried that many in the audience might not really know the difference between viruses, bacteria, and all the rest of Mother Nature's children. And that if folks didn't understand the difference, they wouldn't appreciate how truly unique she is.

"Viruses cannot reproduce by themselves. Instead, they invade a cell and hijack its machinery to make more viruses. In fact," she said with a sniff, "viruses are not even proper organisms. They are little more than bands of rogue genes traveling in a tiny capsule."

I pointed out that although viruses are mainly famous for causing disease—from polio to AIDS—some of them do merit a place in the *Kama Sutra*. For example, the reason that humans must make a new flu vaccine each year is that influenza viruses sometimes have sex, getting new genes that help them sidestep the human immune system. All the same, if Miss *Philodina* were a virus—or a bacterium—no one would be making a fuss about her being asexual. The problem is that, just like any mammal or bird, Miss *Philodina* is a eukaryote.

Unlike bacteria, we eukaryotes keep our genes sequestered in a special place—that is, a nucleus. Eukaryotes come in many shapes and sizes—some have only one cell, others, like bdelloid rotifers and humans, have lots of cells. But for all this diversity, eukaryotes are properly puritanical when it comes to sex. If the mood takes them, bacteria and viruses have lots of different ways to mix genes. Eukaryotes have only one. And when scientists talk of sex being essential, it's eukaryotic sex they mean.

In eukaryotic sex, you get half your genes from your mother and half from your father. But which half? That's decided through the lottery known as meiosis. Say you're playing a game with two decks of fifty-two cards. Each card represents a chromosome—a string of genes. The only rule is that each of your offspring gets one complete deck. It doesn't matter whether they get the queen of spades you got from your father and the jack of diamonds you got from your mother. It doesn't even matter if you cut up both aces of spades and stick them back together again as jumbles of the originals. Indeed, genetic cutting and pasting of this sort is an integral part of eukaryotic sex. It is an internal shuffling of each chromosome, and it is called recombination. Thus, at the end of meiosis, each sperm and each egg carries one complete but unique mixture of genes—a complete but shuffled deck. When egg and sperm from two individuals fuse, they therefore produce a new gene combination. Bdelloid rotifers haven't seen a new gene combination in years—eighty-five million years.

"Eighty-five million years without meiosis. Without genetic exchange. Without men," said Miss *Philodina*. "For eighty-five million years, we've done nothing but clone, and we're jolly proud of it. What's more, we think everyone should follow our example."

She looked all set to go on advocating abstinence, but the belligerent ram in the front leapt to his hooves. He had fixed to his fleece a large badge exhorting everyone to "Save Our Sex." He didn't believe for a moment that bdelloid rotifers are genuine ancient asexuals—or, indeed, that any ancient asexuals exist at all.

Punctuating his speech with vigorous nods of his woolly head, he bleated: "Baaaa. Miss *Philodina*, you claim to be the descendant of a militant virgin who got rid of men and abandoned sex millions of years ago. If what you say is true," he paused to emphasize his skepticism, "it would be sensational. Baaaa. Perhaps you are not aware that other organisms have made this claim—and that it has never stood up to scrutiny."

The ram clearly knew what was at stake. If ancient asexuals really walk the planet, the ramifications—here I bleated respectfully—are dramatic. Again, to spell it out: if they can do without sex or men, maybe the rest of us can too. So you can see why he tried so hard to paint her as a fraud.

"Some of you will remember the case of the chaetonotid gastrotrichs—microscopic animals much like you, Miss *Philodina*, who also live in puddles and mosses," he said. "Like you, they claimed to be ancient asexuals. But when scientists looked at them more closely, they were caught making sperm, an activity not exactly consistent with asexuality, I think you'll agree.

"And who could forget the aphids of the Tramini tribe? Baaaa. What liars! Those fat little insects, too, said they were ancient asexuals. Another hoax! Genetic tests showed they weren't as virtuous as they pretended. And sure enough, scientists found that they kept their males hidden among the roots of the weeds where they live.

"What can we learn from this?" He paused dramatically. "Besides the bdelloid rotifers, a handful of other groups still profess to be ancient asexuals. The darwinulid ostracods—a score of species of small freshwater shellfish—claim to have been without sex for a hundred million years. Baaaa. Certain families of oribatid mites also insist that they got rid of men aeons ago. And there are other self-styled supercelibates: shrimp that live in Old World salt flats, two species of North American fern, and a species of clam. But the evidence for these claims is flimsy, to say the least.

"I put it to you," the ram thundered, "that all claims of ancient asexuality will turn out to be bogus, the supposed celibates unfrocked! Baaaaaaa. Miss *Philodina*, your chastity is a sham! Like the others before you, you are hiding males, and sooner or later you will be exposed!"

As the ram sat down, the audience burst into applause.

I had to say, he was on the mark. Over the years, various organisms *have* claimed to be ancient asexuals; and many of them have indeed been exposed as imposters. Until now, all claims of ancient asexuality have rested on negative evidence—mainly that no one has ever seen a male of that species. Negative claims are weak and easy to dismiss. After all, biology is full of species where the males and females look so different from each other that for decades they were not recognized as being each other's other halves.

Well, I thought the audience was ready to tear up the seats. I must take my hat off to Miss *Philodina*. She kept her cool—and stunned the audience by producing strong evidence that the bdelloid rotifers are not charlatans but genuine ancient asexuals. To everyone's dismay, she made a credible case that she and her foremothers have indeed managed to live without men or meiosis for millions of years. They are wholly virgin, purer than pure, the nun's nuns, ultimate maidens, poster children for abstinence.

Her proof rests on the fact that cloning for millions of years has dramatic effects on the way that genes evolve. "Being asexual for generations leaves an unmistakable mark, a molecular tattoo on your genes," she said smugly. "If you always clone, there's only one source of genetic novelty, only one thing that could cause my genes to differ from my mother's, grandmother's, or great-great-great grandmother's: mutation."

I reminded everyone that mutations are nothing more than sporadic mistakes made by the cell's genetic copying mechanism.

Miss *Philodina* went on, "Let's go back to my ancestor of eighty-five million years ago, the last child of original sin in my family. Say she inherited two copies of a gene for, I don't know, wheel number. One copy came from her father, the other from her mother. And for the sake of argument, let's say they were identical. But now it's eighty-five million years later. Since a bdelloid lives about three weeks, that's about 1.5 billion bdelloid rotifer generations. So you would expect that my two copies of the wheel number gene would be extremely different from each other. Each will have accumulated different mutations."

The best way to understand this process is through an analogy. Imagine an ancient manuscript had been copied again and again by monks in two distant and lonely monasteries. If each new version of the manuscript is copied from the previous one, more and more mistakes will creep in. And unless the monks are telepathic from scriptorium to scriptorium, the mistakes they make will be different. As time goes by, the manuscripts owned by the two monasteries will diverge more and more. In contrast, if you were having sex, it would be as if the monks in the two monasteries were regularly copying from each other, as well as frequently replacing their versions with manuscripts from monasterics elsewhere. The communication between all the monks would ensure that the manuscripts resembled each other closely.

Extensive divergence between the two copies of any given gene, Miss *Philodina* explained, is the molecular stamp of ancient asexuality.

The homing pigeon, his wings twitching with such excitement that he accidentally lifted off, shouted from the air above his perch, "But an ancient text copied 1.5 billion times by two independent groups would change beyond recognition! I don't believe these patterns are detectable!"

"Identifying the two copies of a gene can certainly be difficult. But luckily, in our case, they hadn't changed beyond recognition," said Miss *Philodina*. And then she played her trump card. Triumphantly spinning her wheels in the ram's direction, she put an end to the accusations of imposture, revealing that genetic tests had shown that bdelloid rotifers have the predicted pattern of divergence. Brandishing a copy of *Science* magazine, Miss *Philodina* said with a bounce, "The evidence is conclusive. We bdelloids are celibate. Male bdelloid rotifers do not exist."

The radical feminists at the back greeted this news with a rousing chorus of "That's all right, that's OK, nobody needs them anyway!"

The rest of the audience, however, didn't seem at all happy. As I looked around the studio, I saw long face after long face, and the room hummed with angry murmuring. No one could any longer dispute Miss *Philodina*'s ancient asexual credentials, so instead the crowd became abusive, insinuating the bdelloid rotifers' success was a momentary aberration and that they were surely heading for extinction like other asexuals. A python coiled in a corner raised a large placard that read "The Bdelloids Are Bdoomed" and hissed menacingly, "You're going extinct, you spineless spinster, you're going extinct."

"In the long run we're all going extinct," said Miss *Philodina* tartly. "Sex won't save you from extinction! The dinosaurs had rampant sex, and look what happened to them. You can have sex till you're blue in the face, but if your habitat vanishes, it's you and the dodo. Asexuals—"

The pocket mouse bravely interrupted: "But surely if you don't have sex, you can't adapt to the future? If you can't adapt to the future, you haven't got a future."

"Who says asexuals can't adapt?" Miss *Philodina* spluttered. "I'll have you know that the bdelloid rotifers are one of the most versatile groups on earth. We make up a sisterhood of more than 360 species. We live in the moss, damp soil, funeral urns, gutters, and puddles of seven continents. You'll find us in the wastes of Antarctica and the jungles of Sumatra. We live in sulfurous hot springs and in the purest dew. Compare that with our distant cousins, the seisonid rotifers. They've always had sex, and a fat lot of good it's done them. There are only two species, and they both live on the bodies of one type of shrimp. Call that evolutionary success? Humph. I call it a miserable failure."

I stepped in to stop all the snarling. "The thing to focus on," I said, "is how exceptional the bdelloid rotifers are. They are the only asexual group with lots of species. After eighty-five million years, nobody rational could suppose their extinction is looming. But most asexuals don't survive. Understanding why they don't—and how the bdelloids have—can give us important clues as to why we need sex."

Hallelujah! I'd managed to get the audience back to the matter at hand. I explained that there are more than twenty theories that purport to explain the role of sex, and I briefly summarized the three front-runners, popularly known as Muller's ratchet, Kondrashov's hatchet, and the Red Queen. According to both the ratchet and the hatchet, asexuals are driven extinct by the accumulation of harmful mutations—in other words, asexuals eventually die of genetic diseases. The Red Queen, in contrast, invokes a more traditional horseman of the apocalypse: pestilence, also known as infectious disease.

Moby the puffer fish splashed straight into the subject of harmful mutations: "Miss *Philodina*, without sex, how can asexuals get rid of harmful mutations? And if you'll pardon my saying so, your wheels are looking a little wonky—my eyes aren't too good, so perhaps it's just the light, but the one on the left actually looks *square*. I bet it's those mutations you were saying you've accumulated."

Miss *Philodina* hit right back. "I may be square, but my wheels are not." She sounded confident enough, but I swear I saw her spinning them, just to check. "And if you'll pardon *my* saying so, *dear* puffer fish," she went on, "mutations are greatly overrated as an evolutionary force. Geneticists think mutations are bad because their methods are so crude. All they can see are the bad mutations. *Obviously*, having no head is bad for you. And if you're a fly, having no wings isn't much good. For a start, you'd have to be called a 'walk.' But, in fact, most mutations are neutral. They have no effect. They change the DNA sequence of a gene, sure. But they don't affect the information. It's like switching from the English spelling of a word to the American spelling. "P–l–o–u–g–h" and "p–l–o–w" look different on paper, but they mean the same thing and sound the same when said aloud."

Well, well, well. Given all her mutations, I guess I should have known that Miss *Philodina* would be a neutralist—that she would subscribe to a controversial, even radical school of thought that holds that most mutations are neither helpful nor harmful, just irrelevant. I couldn't let her get away with it, though. First of all, there's a vigorous debate going on over whether most mutations are neutral. And second of all, it's generally agreed that when a mutation does have an effect, the effect is usually bad: small random changes are likely to harm, not help. Or to put it more starkly, many different mutations will kill you or make you sick, but none guarantees success in life. Which brings us to the ratchet and the hatchet.

According to Muller's ratchet (named for its inventor, the geneticist Hermann Muller, who won the Nobel prize for demonstrating that X rays cause mutations), asexuals are evolutionarily short-lived because, over time, the number of harmful mutations they carry will irrevocably and inevitably ratchet upward. Imagine a population that has just become asexual. For the sake of argument, imagine that all members of the population are free of harmful mutations. Over time, copying errors will lead to mutations among their descendants, and gradually the population will consist of individuals who carry several mutations. Then one day the last mutation-free individual will fail to leave children, and the ratchet will have clicked forward one notch. This process continues until eventually all the individuals are so sick that they die and the population goes extinct. Sexuals avoid this fate because the shuffling of genes in each generation produces individuals who carry few mutations.

Muller's ratchet is an elegant idea. But it works only if a number of assumptions are met. The most important of these is that the asexual population is small. In large populations, you see, there may always remain some individuals bearing few mutations. Kondrashov's hatchet (also named for its inventor, a Russian geneticist), however, is another story: it holds regardless of population size.

Suppose there's a threshold number of slightly harmful mutations that any individual can carry. Above that threshold, the hatchet falls—and you're dead. In a population that has sex, the shuffling of genes creates some lucky creatures with few harmful mutations. But it also creates some unlucky ones with many. The unlucky ones fall under the hatchet, taking their mutations to the grave. This quickly and efficiently purges the population of harmful mutations. Asexuals, however, have no such recourse. Far more asexual individuals will cross the threshold of having one bad mutation too many. According to the theory, if the harmful mutation rate is high enough, there is no way to survive without sex.

I summed up, saying, "Harmful mutations could be the reason that most asexuals go extinct. At the moment, we can't measure mutation rates directly, so we can't yet settle the argument. If it turns out that mutation rates among asexuals are generally low, then mutations are not the reason after all. But if it turns out that mutation rates are generally quite high, then it follows that the improbable persistence of the bdelloid rotifers must be due to their having evolved some way to reduce the mutation rate that has eluded other asexuals. Perhaps—"

Suddenly, a mournful voice cut in, "Ah, but Miss *Philodina*, what about infectious diseases?" The screen on the wall had lit up again, and a face appeared. Grotesque, with huge scything mandibles, it looked like an escapee from one of my nightmares. The creature continued: "I am a nameless worker ant of the species *Atta colombica*. I hail from a thriving colony near the Panama Canal; we number more than two million. Long ago, long before humans were a twinkle in Mother Nature's eye, my ancestors invented agriculture, and we have been proudly farming ever since. Although ants of other species farm livestock such as aphids, we farm fungus. Why? For the same reason humans farm wheat or rice. We farm it to eat it—and we cannot live without it. So we cut leaves and flowers from all sorts of different plants to make compost for the fungus to grow on; we manure the fungus with our excretions; we weed the garden where the fungus grows; we prune the fungus to make it more productive; and we try to protect the fungus from pests. We live in constant fear, however, of an event that would jeopardize the entire colony: an outbreak of *Escovopsis*."

The ant shuddered down to the tips of her antennae. She continued: "*Escovopsis* is a virulent disease of the fungus, and if it breaks out, it will destroy the whole garden. This brings me to my question. Our fungus is also an ancient asexual. Not quite as ancient as you, Miss *Philodina*, but it is still thought to be a good twenty-three million years old. We propagate the fungus clonally, and when a new queen leaves her natal nest to start her own colony, she takes some cultivars of fungus with her, packed in special pockets in her throat. This means our fungus gardens are like modern human crops: they are monocultures, whole fields that are genetically identical. We think this is why they are particularly vulnerable to disease. We've heard that susceptibility to disease has a genetic component, so a disease that starts in a monoculture will, pardon the pun, have a field day, destroying everything. One reason sex may be an advantage is because the shuffling of genes gives an edge in the perennial battle against disease. So, Miss *Philodina*, how do you cope with this problem?"

Well, the ant had made an excellent point and brought us right to the third theory of why we need sex, the Red Queen.

As the ant said, susceptibility to infectious disease—or more generally, to parasites, whether viruses, bacteria, fungi, or other nasties—typically has a genetic component. Since asexuals keep the same genes (give or take a mutation or two) from one generation to the next, parasites can easily evolve to infiltrate their defenses, annihilating clones. In contrast, sex, by mixing up genes, prevents parasites from becoming too well adapted to their hosts. Sex is an advantage *because* it breaks up gene combinations: it creates the genetic version of a moving target. With each act of sex, the parasites have to start again from square one. The name of the theory, the Red Queen, comes from *Through the Looking-Glass*. Remember? The Red Queen says to Alice, "Now, *here*, you see, it takes all the running *you* can do, to keep in the same place." In other words, you have to change to stay where you are.

Before Miss *Philodina* could marshal her arguments, the supercilious armadillo rose to his feet. The lights glinting off the back of his shiny carapace, he said: "If I might speak." He waved a front leg dismissively. "I am a nine-banded armadillo, and I believe I am in an unusual position. Armadillos are rare, yes—I might even be so bold as to say unique—among mammals: we routinely engage in both sexual and asexual reproduction. When boy armadillo meets girl armadillo"—he sniggered—"I'll spare you the details, but as I'm sure you know, armadillos are unusual in the length of their equipment. It has to be reinforced with special fibers, so you can just call me Mr. Big. Anyway, egg fuses with sperm, and that cell then splits and splits again to give four genetically identical embryos. Thus, I am genetically different from my parents, yes, but a clone of my brothers.

"But I digress. The real point of the Red Queen is this: sex is an advantage because it makes you rare. Monocultures are vulnerable to disease because all the individuals are the same clone. A field full of different clones would not be vulnerable, yes? The disease would not be able to sweep through and infect everybody.

"Ironically, successful clones are the cause of their own demise. As a clone increases in frequency, it becomes more vulnerable to disease, yes, for two reasons: first, a disease can spread easily between members of the same clone; and second, as it becomes more common, the disease has more chances to evolve to infiltrate the target."

Everyone started whispering. The armadillo stamped his feet for attention. As the audience calmed down, he went on: "How do you avoid being common if you're a clone? Easy. You go somewhere else. You see, when a clone arrives in a new area, yes, she will have all the benefits of being unique. Assuming she doesn't carry her parasites with her and assuming the diseases she's left behind are worse than the diseases in the new place, she should be able to go on without sex for as long as she can keep on the move."

"Oh, of course! Our fungus travels!" said the ant.

"Yes, and in conditions that reduce the chances that *Escovopsis* goes with it," said the armadillo. "I don't know, but I'd guess that the cultivar of fungus the queen takes with her is chosen carefully, yes, and perhaps even disinfected."

"Oh, *Escovopsis* never travels with the queen. We know that," said the ant.

"So then the question to Miss *Philodina* is, do you travel? Is this how you outwit the Red Queen?" said the armadillo grandly.

I began to feel superfluous. But I guess this is a subject where everyone wants to put in their two cents. As for Miss *Philodina*, well, she looked livid. And she sounded it, too, replying snappishly, "The answer's in my forthcoming book, *The Seven Habits of Highly Successful Asexuals*, which is, I'll have you know, under embargo. But since you've as good as told the secret, I might as well spill the beans.

"I believe that travel is indeed the secret of our success. We bdelloids travel in both space and time. Of course, we can't travel backward in time: nobody can do that. But we can go forward. We have a trick called anhydrobiosis. It's a state of suspended animation. Essentially, we dry up and blow away."

Someone in the audience sneered that the old crone was already dried up.

Miss *Philodina* pretended not to hear. "It's risky. Anhydrobiosis is difficult," she said. "Many bdelloids never recover. But if you do survive, you come back to life in a new place and time, healthier and happier than before."

I made my last stab at taking back my show. Drying up and blowing away may not—by itself—be enough to make a successful career of asexuality, I pointed out. After all, other organisms have anhydrobiosis, and they are not ancient asexuals. I imagine, though, that this strange ability to travel in space and time is an important factor in the bdelloids' success. Long-term success as an asexual is difficult—and almost certainly requires several quirks of fate.

By this point, I was exhausted. But the show had made a crucial point: although we don't have a definitive answer, it looks like we need sex to stay healthy. Shuffling genes can help us evade parasites and reduces the impact of harmful mutations. In short, sex enables us to survive.

I concluded the show with a double warning: girls in most species would still be unwise to get rid of males altogether (this was greeted by boos from the radical feminists), and males—especially mammals—should avoid becoming complacent: "You male mammals may be busy congratulating yourselves. Of all the animals, mammals are the only ones where the asexual reproduction of adults is unknown. Mammalian males have evolved to be indispensable—genetically at least. The cloning of adult mammals is not possible without heavy and, for the moment, unreliable technological hocus-pocus.

"So, men, you're safe for now. But if you don't want to be abolished, let me give you some advice. Asexuality is particularly attractive to girls in species where males are lazy and never give a hand with the child care. The benefits of asexuality are greatly reduced if males help out."

Whew. I'd made it through without a riot's breaking out or the plug's being pulled. I thanked everyone for their questions and finished by calling for a big hand for Miss *Philodina roseola:* "Long live the bdelloid rotifers!" The audience burst into applause as I recited my tag line, "Predators, please remember that it is forbidden to eat guests as they leave. And all of you, join me again next week when we'll put another deviant *Under the Microscope!*"

# POSTSCRIPT

There's time for a last question—one for the road, so to speak. And with apologies to all whose questions I haven't answered, I thought I'd pose one that's often on my own mind. We've seen that sex is central to evolution, that it generates fabulous diversity, that despite the trouble it causes, it's something most of us can't live without, that abstinence almost always leads to extinction. But how did sex begin?

Alas, we may never know for sure. After all, some sort of genetic exchange probably started soon after life appeared about four billion years ago—and looking that far back in time is an uncertain business at best. But there are lots of wacky ideas. Let's take a quick glance at some of them.

Microbes not that different from modern bacteria seem to have evolved shortly after the origin of life, so it's tempting to imagine that gene exchange among these primordial beasties proceeded much as it does now. But why did they start swapping genes in the first place? One idea is that gene exchange facilitated the repair of damaged DNA: an intact string of DNA received from a partner could perhaps be used to replace or repair genes that had broken. A second, more exotic idea is that sex was simply infectious. In other words, it arose because a segment of DNA promoted gene exchange in order to spread itself through the population. To use an analogy, it's as though the common cold caused humans to be promiscuous—an effect that would clearly enhance its transmission. But although such a scenario sounds crazy, it may not be. One reason a modern bacterium will be moved to have sex is because it's become infected with a particular segment of DNA known as the F plasmid. An individual who's got the F plasmid is then driven to mate with an individual who hasn't, and so spreads the sex habit around.

However speculative the origins of bacterial sex, though, the ideas look like a mighty edifice compared with how little we know about the origins of the sort of sex that humans, birds, bees, fleas, green algae, and other eukaryotes conduct. Remember: eukaryotic sex is a complicated process requiring that each parent donate one complete set of his or her genes. Probably it evolved only once. But exactly how or why is a deep mystery. Some say it arose as a result of cannibalism—one cell eats another and then collects its DNA. Others plump for DNA repair. And still others argue that this sort of sex, too, originated as a disease, with infectious genetic elements promoting their own spread.

I leave you with these thoughts. I hope that having seen the prodigious variety of sexual practices out there, you'll be more tolerant of the predilections of others. Speaking for myself, my years as a sex adviser have definitely broadened my horizons; I now think that many more things are normal. Indeed, I must confess I envy some of you. (Who? Sorry, that's a secret.) In any case, I hope I've helped you to put your problems in perspective—and above all to relax and have fun.

Wishing all of you—except the feisty Miss *Philodina*, of course—lots of great sex in the years ahead.

So long!

*Dr. Tatiana*

# NOTES

Notes for each chapter are preceded by a list of the scientific names of species whose common names are given in the text. I have not included generic names, like goldfish, that refer to several species at once.

## Chapter 1: A Sketch of the Battlefield

| | |
|---|---|
| Stick insect | *Necroscia sparaxes* |
| Idaho ground squirrel | *Spermophilus brunneus* |
| Blue milkweed beetle | *Chrysochus cobaltinus* |
| Alfalfa leaf-cutter bee | *Megachile rotundata* |
| Rabbit | *Oryctolagus cuniculus* |
| Gunnison's prairie dog | *Cynomys gunnisoni* |
| Sand lizard | *Lacerta agilis* |
| Slippery dick | *Halichoeres bivattatus* |
| Golden potto | *Arctocebus calabarensis* |
| Dunnock | *Prunella modularis* |
| Red-billed buffalo weaver | *Bubalornis niger* |
| Gorilla | *Gorilla gorilla* |
| Argentine lake duck | *Oxyura vittata* |
| Honeybee | *Apis mellifera* |
| House mouse | *Mus musculus* |
| Fox squirrel | *Sciurus niger* |
| Rat | *Rattus norvegicus* |

## Sick of Sex in India

For sex marathons in stick insects, see Gangrade (1963). For mate guarding in the Idaho ground squirrel, see Sherman (1989); in the blue milkweed beetle, see Dickinson (1995). For the original statement of Bateman's principle and for Bateman's experiments, see Bateman (1948). For extra reasons that females should have as little sex as possible, see Daly (1978); for single mating in the alfalfa leafcutter bee, see Gerber and Klostermeyer (1970). For goldfish being drowned by frogs, see Boarder (1968). For enthusiasm for sex in *Drosophila hydei*, see Markow (1982); for increasing fertility with increasing numbers of mates in *Drosophila melanogaster*, see Pyle and Gromko (1978). For early recognition of problems with Bateman's principle, see Gladstone (1979), Mason (1980), and Dewsbury (1982); for the notion that promiscuous females have "malfunctioned," see Taylor (1967). Parker (1970) was among the first to recognize that females in some insect species may be routinely promiscuous, and the first to explore the consequences for males; however, he did not consider that sex could be beneficial to females. For early recognition that females routinely benefit from multiple mating, see Ridley (1988). For higher rates of conception with multiple mating in rabbits, see Beatty (1960); in Gunnison's prairie dogs, see Hoogland (1998); in sand lizards, see Olsson and Madsen (2001); in slippery dicks, see Petersen (1991).

## Spooked in Gabon

For what little is known of the natural history of the golden potto, see Nowak (1999), pages 495–96; for the golden potto's penis, see Hill (1953), pages 164–75; for general descriptions of primate penises, see Dixson (1998), chapter 9; for how the human penis compares, see Short (1979). For scouring in *Calopteryx maculata*, see Waage (1979); for persuasion in *Calopteryx haemorrhoidalis asturica*, see Córdoba-Aguilar (1999); for genitalia of *Olceclostera seraphica*, see Eberhard (1985), page 165; of termites, see Eberhard (1985), pages 126–27. For sperm sealing in ghost spider crabs, see Diesel (1990); for pecking in dunnocks, see Davies (1983). For the pseudophallus and copulation in buffalo weavers, see Winterbottom et al. (1999 and 2001). For penis complexity and female promiscuity in insects, see Arnqvist (1998); in primates, see Dixson (1998), chapter 9. For anecdotal reports of females having more orgasms in promiscuous species, see Dixson (1998), pages 131–36; note, however, that the function of spines on mammalian penises remains controversial. For the long and prickly penis of the lake duck, see McCracken (2000); for group structure in gorillas, see Robbins (1999).

## Perplexed in Cloverhill

For the violent death of the male honeybee, see Gary (1963); for male numbers in swarms, see Page (1980); for number of queen flights, see Page (1986). David Tarpy told me of the number of mates of a honeybee queen. For benefits of multiple mating in honeybees, see Page (1980), Page (1986), and Tarpy and Page (2001); for the eating alive of infertile males, see Woyke (1963). Many thanks to

David Tarpy, Robert Page, and J. Woyke for telling me of the plug-removal structure on the tip of the male phallus. For examples of chastity plugs in bats, see Matthews (1941); in worms, see Barker (1994); in snakes, see Devine (1975); in spiders, see Robinson (1982); in butterflies, see Dickinson and Rutowski (1989); in fruit flies, see Polak et al. (1998); in guinea pigs, see Martan and Shepherd (1976); in squirrels, see Koprowski (1992); in chimpanzees, see Tinklepaugh (1930). For plugs in rodents (including the tough plug of the house mouse), see Voss (1979). For plug removal by female fox squirrels, see Koprowski (1992); for gymnastics in the rat penis, see Wallach and Hart (1983).

## Chapter 2: The Expense Is Damnable

| | |
|---|---|
| Splendid fairy wren | *Malurus splendens* |
| Yellow dung fly | *Scatophaga stercoraria* |
| Seaweed pipefish | *Syngnathus schlegeli* |
| Honeybee | *Apis mellifera* |
| Rabbit | *Oryctolagus cuniculus* |
| Roundworm | *Caenorhabditis elegans* |
| Bulb mite | *Rhizoglyphus robini* |
| Painter's frog | *Discoglossus pictus* |
| Jack-in-the-pulpit | *Arisaema triphyllum* |
| Lemon tetra | *Hyphessobrycon pulchripinnis* |
| Bluehead wrasse | *Thalassoma bifasciatum* |
| Garter snake | *Thamnophis radix* |
| Zebra finch | *Taeniopygia guttata* |
| Blue crab | *Callinectes sapidus* |
| Sheep | *Ovis aries* |
| Adder | *Vipera berus* |
| Lion | *Panthera leo* |
| Rat | *Rattus norvegicus* |
| Golden hamster | *Mesocricetus auratus* |
| Cactus mouse | *Peromyscus eremicus* |
| Crested tit | *Parus cristatus* |
| Leopard | *Panthera pardus* |
| Tiger | *Panthera tigris* |
| Puma | *Puma concolor* |
| Jaguar | *Panthera onca* |
| Cheetah | *Acinonyx jubatus* |
| Snow leopard | *Panthera uncia* |
| Sand cat | *Felis margarita* |
| Bobcat | *Lynx rufus* |
| Tree ocelot | *Leopardus wiedii* |
| Giant water bug | *Abedus herberti* |
| Long-tailed dance fly | *Rhamphomyia longicauda* |

Scarlet-bodied wasp moth       *Cosmosoma myrodora*
Mormon cricket               *Anabrus simplex*

## Bewildered Down Under

For general biology of splendid fairy wrens, see Russell and Rowley (1996); for their sperm counts, see Tuttle et al. (1996); for feather carrying, see Rowley and Russell (1990). For sperm numbers in humans, see Cohen (1971), and Harvey and May (1989). For pollination and differences in pollen production among fig species, see Kjellberg et al. (2001). For sperm and egg numbers in externally fertilizing fish, see Stockley et al. (1996). For the discovery of sperm competition, see Parker (1970); for the role of sperm competition in the evolution of sperm numbers, see Parker (1990a); for patterns of larger testes size and greater sperm numbers as a result of female promiscuity, see Møller (1989); for the experimental manipulation of sperm competition in yellow dung flies, see Hosken and Ward (2001). For sperm counts in the seaweed pipefish, see Watanabe et al. (2000). For early observations of high sperm death, see van Leeuwenhoek, cited in Cohen (1971). For the hypothesis that female reproductive tracts are a kind of obstacle course and for an overview of the hazards in humans, see Cohen (1971) and Birkhead et al. (1993); for sperm digestion, see Tompa (1984) and Michiels (1998); for ejection, see Morton and Glover (1974); for other types of sperm removal, see Hanlon and Messenger (1996), page 117. For sperm numbers and storage in honeybees, see Laidlaw and Page (1984). For acidity in the human vagina, see Masters and Johnson (1966), pages 88–100; Roger Short told me of lemons making good contraceptives. For white blood cells amassing at the cervix in rabbits, see Phillips and Mahler (1977); in humans, see Pandya and Cohen (1985) and Barratt et al. (1990). For sperm numbers reaching the oviducts of rabbits, see Cohen and Tyler (1980); of humans, see Ahlgren (1975). For infertile human sperm counts, see MacLeod and Gold (1951). For sperm numbers and sperm survival in the rabbit, see Morton and Glover (1974). For substances in semen that suppress the female immune system, see Mann and Lutwak-Mann (1981). For genetic testing of splendid fairy wren chicks, see Brooker et al. (1990).

## Waiting for Sperm in Ohio

For sperm size in *Drosophila bifurca*, see Pitnick et al. (1995); in humans, see Dixson (1998), page 228. For evidence that sperm tend to be smaller and simpler in externally fertilizing species, see Franzén (1977). For tandem sperm in American opossums, see Moore (1992); in water beetles, see Mackie and Walker (1974); in millipedes, see Jamieson et al. (1999), page 281; in firebrats, see Dallai and Afzelius (1984); in marine snails, see Afzelius and Dallai (1983). For hooked sperm in koalas, see Hughes (1965); in rodents, see Roldan et al. (1992); in crickets, see Jamieson et al. (1999), chapter 9. For sperm that resemble flat discs in the protura, see Jamieson et al. (1999), page 73; Catherine wheels in crayfish, see

Moses (1961); corkscrews in land snails, see Tompa (1984), page 120. For bearded sperm in termites, see Jamieson et al. (1999), page 134; for amoeboid sperm in roundworms, see Ward and Carrel (1979). For spermatophores in giant octopus, see Mann et al. (1966). For large-sperm advantage in roundworms, see LaMunyon and Ward (1998); in bulb mites, see Radwan (1996). For the general relationship between larger sperm and female promiscuity, see Gomendio and Roldan (1991) and Dixson (1993). For declining sperm numbers with increasing sperm size in fruit flies, see Pitnick (1996). For giant sperm in featherwing beetles, see Dybas and Dybas (1981) and Taylor et al. (1982); in back-swimming beetles, see Afzelius et al. (1976); in ostracods, see Lowndes (1935) and Gupta (1968); in ticks, see Rothschild (1961); in *Hedleyella falconeri*, see Thompson (1973); in the painter's frog, see Afzelius et al. (1976); in fruit flies, see Pitnick et al. (1995). Peter Henderson told me of fighting ostracod sperm; see also Lowndes (1935). For big sperm having nothing to do with big eggs in fish, see Stockley et al. (1996); in fruit flies, compare egg sizes given in Atkinson (1979) with sperm sizes given in Pitnick et al. (1995). For the notion that giant sperm are a kind of present, see Bressac et al. (1994); for evidence against this, see Karr and Pitnick (1996). For giant sperm as chastity belts, see Ladle and Foster (1992), and for evidence that they may be used this way in featherwing beetles, see Dybas and Dybas (1981). Peter Henderson told me of ostracod sperm having to travel outside the female's body to reach the eggs. For sperm size in *Drosophila hydei*, see Pitnick et al. (1995); for female mating habits and sperm mixing in this species, see Markow (1985). For general costs of making giant sperm in fruit flies, see Pitnick et al. (1995); for delay in sperm production and testes size in *Drosophila bifurca*, see Pitnick et al. (1995); for *Drosophila pachea* spending the first half of his adult life unable to reproduce, see Pitnick et al. (1995). Scott Pitnick told me of the lifespan of *Drosophila bifurca*.

### Dried Up in London

For sterility in *Drosophila melanogaster* males, see Prowse and Partridge (1997); for Bateman's ideas on sperm being unlimited, see Bateman (1948). For reviews of sperm limitation in marine organisms, see Levitan and Petersen (1995) and Yund (2000); for sponges spewing sperm, see Reiswig (1970). For pollinators preferring to eat pollen, see Bierzychudek (1987); for an overview of pollen limitation, see Burd (1994); for pollen limitation in jack-in-the-pulpit, see Bierzychudek (1981). For sperm limitation in the lemon tetra, see Nakatsuru and Kramer (1982); in the bluehead wrasse, see Shapiro et al. (1994) and Warner et al. (1995). For early comments on the costs of ejaculates, see Gladstone (1979), Baylis (1981), and Dewsbury (1982). For sexual exhaustion in garter snakes, see Ross and Crews (1977); for sperm depletion in zebra finches, see Birkhead et al. (1995). For sperm depletion in blue crabs, see Jivoff (1997); in rams, see Synnott et al. (1981). For supposed sperm reserves in rams and humans, see Møller (1989). For adders losing weight as a result of producing sperm, see Olsson et al. (1997). For sperm limitation, sperm production, and unfertilized eggs in *C. elegans,* see Ward and Carrel (1979); for the disadvantage of making more sperm, see Hodgkin and Barnes (1991) and Barker (1992). For sperm limitation in sea slugs, see, for example, *Onchidoris fusca* in Hadfield (1963) and *Hermissenda crassicornis* in Rutowski (1983); in aquatic snails, see Jarne et al. (1993); in sea hares, see Yusa (1996). For sperm depletion in *Dugesia gonocephala,* see Vreys and Michiels (1998); for *Navanax inermis* preferring the female role over the male role, see Leonard and Lukowiak (1985); for banana slugs gnawing off their own phallus, see Mead (1942). For temporary reductions in fertility in *Drosophila melanogaster,* see Markow et al. (1978); for permanent sterility, see Prowse and Partridge (1997); for female *Drosophila* preferring virgins, see Markow et al. (1978).

### Sex Machines Aren't Us in the Serengeti

For copulation rates in lions, see Bygott et al. (1979) and Estes (1991), page 376. The classification of sex mania was developed by Dr. Tatiana. For females needing stimulation to become pregnant, see Wilson et al (1965) (rats); Lanier et al. (1975) (golden hamsters); Dewsbury and Estep (1975) (cactus mice). For induced ovulation in rabbits, ferrets, and domestic cats, and for pregnancy failure in unstimulated rats, see Wilson et al. (1965); for indications that lions may actually ovulate spontaneously, see Schmidt et al. (1979). For evidence that lionesses are least likely to conceive after a takeover, see Packer and Pusey (1983). For reduced female receptiveness after vigorous sex in golden hamsters, see Huck and Lisk (1986); in rats, see Lanier et al. (1979); in crested tits, see Lens et al. (1997). For paternity data in lions, see Packer et al. (1991). For copulation rates in the cat family, see Eaton (1978). For copulation behavior in giant water bugs, see Smith (1979).

### Quasimoda in Delaware

For the flying-saucer look in long-tailed dance flies, see Funk and Tallamy (2000). For males with small gifts not being allowed to copulate for long, see Thornhill (1976) and Svensson et al. (1990); for presents in the hunting spider, see Austad and Thornhill (1986) and Lang (1996). For a general review of gift giving in insects, see Vahed (1998). For anal feeding in the tropical cockroach, see Schal and Bell (1982). For spider repellent in *Utetheisa ornatrix,* see Gonzalez et al. (1999); in the scarlet-bodied wasp moth, see Conner et al. (2000). For balloons in balloon flies, see Aldrich and Turley (1899) and Kessel (1955). For reduced circumstances in Mormon crickets, see Gwynne (1981); for reduced circumstances in butterflies, see Vahed (1998).

### Chapter 3: Fruits of Knowledge

Bronze-winged jacana                 *Metopidius indicus*
Greater rhea                         *Rhea americana*

| Red-winged blackbird | Agelaius phoeniceus |
|---|---|
| Reed bunting | Emberiza schoeniclus |
| Giant honeybee | Apis dorsata |
| Orange-rumped honeyguide | Indicator xanthonotus |
| Field grasshopper | Chorthippus brunneus |
| Green-veined white butterfly | Pieris napi |
| Shiner perch | Cymatogaster aggregata |
| Zebra finch | Taeniopygia guttata |
| Peacock | Pavo cristatus |
| Mallard | Anas platyrhynchos |
| Harlequin-beetle-riding pseudoscorpion | Cordylochernes scorpioides |
| Harlequin beetle | Acrocinus longimanus |
| Chimpanzee | Pan troglodytes |
| Rat | Rattus norvegicus |
| Yellow dung fly | Scatophaga stercoraria |
| Caribbean reef squid | Sepioteuthis sepioidea |
| Farmyard chicken | Gallus gallus domesticus |

## Neglected Househusband in Tamil Nadu

For yelling in the bronze-winged jacana, see Butchart et al. (1999). For child care in the greater rhea, see Handford and Mares (1985); in fish, see Blumer (1979). For females receiving more help in red-winged blackbirds, see Gray (1997); for males withdrawing help from promiscuous female reed buntings, see Dixon et al. (1994). Stuart Butchart told me of the status of paternity testing in bronze-winged jacanas.

## Waxing Suspicious in the Himalayas

For the breeding biology of the orange-rumped honeyguide, see Cronin and Sherman (1976). For increased reproductive success with food consumption in grasshoppers, see Butlin et al. (1987). For benefits to female remating in the green-veined white butterfly, see Kaitala and Wiklund (1994). For the controversy over male feeding of females as a mating ruse versus parental investment, see Vahed (1998).

## Outraged in Baja

For the sex life of shiner perch, see Shaw and Allen (1977) and Darling et al. (1980). For the notion that females remate to hedge against male sterility, see Darling et al. (1980); because they have run out of sperm, see Walker (1980); to promote genetic diversity of their offspring (and for why genetic diversity is unlikely to be an important force in female remating), see Williams (1975), page 129. For the role of genetics in female inclination to remate in Drosophila melanogaster, see Pyle and Gromko (1981); in field crickets, see Solymar and Cade (1990).

242 NOTES

*Feeling Inadequate in Malaysia*

For female preferences in *Cyrtodiopsis dalmanni*, see Wilkinson and Reillo (1994); Andrew Pomiankowski told me of the rampant promiscuity by females in this species. For Darwin on sexual selection, see Darwin (1871 [1981]). For the original statement of Fisher's runaway process, see Fisher (1999), pages 131ff. For good genes driving female choice, see Williams (1966), page 184. For bracelets on zebra finches, see Burley et al. (1982); for females laying an extra clutch for males with red bracelets, see Zann (1994). For improved survival of peacocks whose fathers had good tails, see Petrie (1994); for female mallards laying larger eggs for sexy drakes, see Cunningham and Russell (2000); for females adjusting the testosterone composition of eggs, see Gil et al. (1999). For condition dependence of eyespan in stalk-eyed flies, see David et al. (2000).

*Stranded in Panama*

For general biology of the harlequin-beetle-riding pseudoscorpion, see Zeh and Zeh (1992); for benefits of multiple mating, see Zeh (1997) and Newcomer et al. (1999); for genetic consequences of multiple mating, see Zeh et al. (1997); for female dislike of previous mates, see Zeh et al. (1998). Trivers (1972) suggested that females might choose mates on the basis of genetic compatibility; it is only recently, however, that genetic incompatibilities have been proposed as a cause of female promiscuity—see Zeh and Zeh (1996) and Zeh and Zeh (1997). For genetic interactions and sperm competition in *Callosobruchus maculatus*, see Wilson et al. (1997). For rates of infertility and incompatibility in humans, see Mandelbaum et al. (1987). For the role of the major histocompatibility complex in spontaneous abortions, see Beer et al. (1982), Ober et al. (1998), and Creus et al. (1998). For people smelling the difference between mice that differ only at the MHC, see Gilbert et al. (1986); for mice smelling the difference between mice, see Beauchamp et al. (1985). For mixed results regarding whether humans preferentially mate with people who differ at the MHC, see Edwards and Hedrick (1998); for smelly T-shirt experiments (including the reversal of preferences among women on the Pill), see Wedekind et al. (1995) and Wedekind and Füri (1997).

*Mind Boggling and Eyes Popping in the Ivory Coast*

For promiscuity in female chimpanzees, see Goodall (1986), pages 443 and 454. For the notion that female chimpanzees are promiscuous in order to promote sperm competition, see Harvey and May (1989) and Keller and Reeve (1995); for detailed analysis of the sperm competition hypothesis, see Curtsinger (1991) and Keller and Reeve (1995). For variation in outcomes of sperm competition, see Simmons and Siva-Jothy (1998); for sperm competition results in different halves of the rat's reproductive tract, see Spinka (1988). For mitochondria causing infertility, see Folgerø et al. (1993), Bourgeron (2000), and Wei and Kao (2000). For sperm traits that are heritable, see references in Keller and Reeve

(1995). For the massive obfuscation hypothesis of female promiscuity, see Hrdy (1979). For infanticide in chimpanzees, see Nishida and Kawanaka (1985).

## The Dandy on the Cowpat

For a general discussion of the notion that females may actively select sperm, see Eberhard (1996); for a critique, see Birkhead (1998). For sperm rejection in the reef squid, see Hanlon and Messenger (1996), page 117; in the chicken, see Pizzari and Birkhead (2000). For an overview of comb-jelly biology (including *Beroë* zipping its mouth shut), see Margulis and Schwarz (1998), pages 224–27; for reproductive biology of comb jellies, see Harbison and Miller (1986); for sperm choice in *Beroë ovata*, see Carré and Sardet (1984) and Carré et al. (1991). For polyspermy in sharks, see Wourms (1977). For patterns of sperm use in the mallard, see Cunningham and Cheng (1999). For claims that the yellow dung fly chooses particular sperm, see Otronen et al. (1997) and Ward (1998); for counterclaims, see Simmons et al. (1996). For sperm displacement and copulation duration in the yellow dung fly, see Parker and Simmons (1994).

## Summary

For the notion that female promiscuity is a malfunction, see Taylor (1967); a response to harassment, see Alcock et al. (1977) and Parker (1979). For females getting clogged up in *Macrocentrus ancylivorus*, see Flanders (1945).

## Chapter 4: Swords or Pistols

| | |
|---|---|
| Fig wasp | *Idarnes* (most parasitic fig wasps have not been described in enough detail to have a full Latin name) |
| Gladiator frog | *Hyla rosenbergi* |
| African elephant | *Loxodonta africana* |
| Northern elephant seal | *Mirounga angustirostris* |
| Burmese jungle fowl | *Gallus gallus spadiceus* |
| Cheetah | *Acinonyx jubatus* |
| Two-spotted spider mite | *Tetranychus urticae* |
| European bedbug | *Cimex lectularius* |
| Jordan salamander | *Plethodon jordani* |
| Three-spined stickleback | *Gasterosteus aculeatus* |

### Give Peace a Chance in Ribeirão Prêto

For an overview of the biology of parasitic fig wasps, see West et al. (1996); for descriptions of fighting in parasitic fig wasps, see Hamilton (1996), chapter 13. For general biology of fig trees, see Janzen (1979b); I learned of the absence of fig wasps on Kauai during my stay on the island. For biology of the fig pollinator wasps, see Wiebes (1979) and Anstett et al. (1997). For seed loss in fig trees, see Janzen (1979a). For the evolution of winged and wingless males in parasitic fig wasps, see Hamilton (1996), chapter 13, and Cook et al. (1997). For conditions favoring the evolution of fatal fighting, see Enquist and Leimar (1990) and (especially with regard to fig wasps) West et al. (2001). For fighting among annual fishes, see Myers (1952). For the biology of gladiator frogs, see Kluge (1981). For violent orchids, see Romero and Nelson (1986). Stuart West told me of trying to study fighting in fig wasps and of never being there in time.

## Anxious in Amboseli

For green-penis syndrome, and other signs of musth in elephants, see Poole and Moss (1981), Poole (1987), and Poole (1989a); for females preferring large, old males, see Moss (1983); for their summoning bigger bull elephants, see Poole (1989b). For females inciting male competition in northern elephant seals, see Cox and Le Boeuf (1977). For incitation in the Burmese jungle fowl, see Thornhill (1988). For goading in *Zootermopsis nevadensis*, see Shellman-Reeve (1999). For female cheetahs being aroused by male fights, see Caro (1994), page 42. For fighting in the two-spotted spider mite, see Potter et al. (1976). For fighting in boa constrictors, see Tolson (1992); for a general account of the size advantage in fighting, see Andersson (1994), chapter 11; for fighting conventions in stalk-eyed flies, see de la Motte and Burkhardt (1983). For discussion of delayed fusion of the long bones in elephants, see Poole (1994). For testosterone levels of elephants in musth, see Poole (1989a). For elephant conversations and sex-specific vocabularies, see Poole (1994). For escalation of fights in elephants, see Poole (1989a).

## Making Mischief between the Sheets

For claims that male *Xylocoris maculipennis* inseminate one another and for details of their genitalia, see Carayon (1974). In a conversation in August 2001, Mike Siva-Jothy told me that experiments on the European bedbug have failed to find evidence that males inject one another. For gonadal hijacking in *Botryllus schlosseri*, see Stoner and Weissman (1996). For cementing in *Moniliformis dubius*, see Abele and Gilchrist (1977). For hypothesized spoiling tactics in African bat bedbugs, see Carayon (1959). For spermatophore trampling and eating and for spermatophore stalagmites, see Proctor (1998); for sabotage in *Plethodon jordani*, see Arnold (1976).

## Want My Eggs Back in Vancouver

For egg theft in sticklebacks, see Wootton (1971), Rohwer (1978), and Mori (1995); for evidence that females prefer nests with eggs, see Ridley and Rechten (1981); for the notion that stickleback eggs may be easier to steal than other fish eggs, see Jones (1998). For general biology of bowerbirds, see Diamond (1986); for bowerbirds trying to steal cameras and socks and for a general description of the artists at work, see Diamond (1988). For vandalism and theft among bowerbirds, see Borgia (1985a) and Borgia and Gore (1986). For evidence that female bowerbirds go for the most lavish displays, see Borgia (1985b).

## Chapter 5: How to Win Even If You're a Loser

| | |
|---|---|
| Sponge louse | *Paracerceis sculpta* |
| Bluegill sunfish | *Lepomis macrochirus* |
| Horseshoe crab | *Limulus polyphemus* |
| Side-blotched lizard | *Uta stansburiana* |
| Peacock | *Pavo cristatus* |
| Bluehead wrasse | *Thalassoma bifasciatum* |
| Southern sea lion | *Otaria byronia* |
| Hammerheaded bat | *Hypsignathus monstrosus* |
| Black grouse | *Tetrao tetrix* |
| Field cricket | *Gryllus integer* |
| Bullfrog | *Rana catesbeiana* |
| Parasitic fly | *Ormia ochracea* |
| Fringe-lipped bat | *Trachops cirrhosus* |
| Little blue heron | *Florida coerulea* |
| Mediterranean house gecko | *Hemidactylus tursicus* |
| Decorated cricket | *Gryllodes supplicans* |
| Marine iguana | *Amblyrhynchus cristatus* |
| Sooty mangabey | *Cercocebus torquatus* |
| Orangutan | *Pongo pygmaeus* |
| Chimpanzee | *Pan troglodytes* |
| Red deer | *Cervus elaphus* |
| Plainfin midshipman fish | *Porichthys notatus* |

### Hoodwinked in the Gulf of California

For the story of Caesar and Pompeia, see Suetonius, page 11; for the origin of the "above suspicion" quotation, see Plutarch, page 860. For general biology of the sponge louse, see Shuster (1991). For a comprehensive overview of sneaking in fish, see Taborsky (1994). For sneaking in the bluegill sunfish, see Gross and Charnov (1980). For sneaking in elderly male damselflies, see Forsyth and Montgomerie (1987); in horseshoe crabs, see Brockmann and Penn (1992). Bryan Danforth kindly told me of the development of *Perdita portalis*. For the genetics of male morphology in sponge lice, see Shuster and Wade (1991); for female preferences, see Shuster (1990). For fights in the sponge louse, see Shuster (1989) and Shuster (1990). For beta males behaving like females, see Shuster (1989); for the proportions of different male types in the population, see Shuster and Wade (1991). For the success of male types in the side-blotched lizard, see Sinervo and Lively (1996) and Zamudio and Sinervo (2000).

### Invisible in Sri Lanka

For groups of young male bluehead wrasse invading territories, see the account in Warner and Schultz (1992) of the natural history of this species. For charges in southern sea lions, see Campagna et al. (1988). For a general discussion of the evolution of leks, see Höglund and Alatalo (1995). For sand castles in *Cyrtocara eucinostomus*, see McKaye (1983) and McKaye et al. (1990). For honking in the hammerheaded bat, see Bradbury (1977). For an example of females preferring larger leks, see Alatalo et al. (1992). For peacocks lekking with their brothers, see Petrie et al. (1999); for black grouse lekking with their brothers, see Höglund et al. (1999). For the composition of lion coalitions, see Packer et al. (1991); for skewed sex ratios, see Packer and Pusey (1987).

## Hoppin' Mad in Texas

For satellite males in luminescent ostracods, see Morin (1986) and Morin and Cohen (1991). For skulking in bullfrogs, see Howard (1978). For female flies being attracted to singing crickets, see Cade (1975), Cade (1979), and Gray and Cade (1999); for genetic variation in crickets' inclination to sing, see Cade (1981). For a general account of predation on signaling males, see Burk (1982); for fringe-lipped bats preying on calling frogs, see Tuttle and Ryan (1981); for little blue herons eating calling crickets, see Bell (1979); for Mediterranean house geckos lying in wait for female decorated crickets, see Sakaluk and Belwood (1984). For the hunting habits of adult female bolas spiders, see Yeargan (1994); of juvenile bolas spiders, see Yeargan and Quate (1996); of male bolas spiders, see Yeargan and Quate (1997). For predation by *Oxybelus exclamans* on parasitic flesh flies, see Peckham and Hook (1980).

## Disgusted in the Galápagos

For masturbation in marine iguanas, see Wikelski and Bäurle (1996). For a general discussion of masturbation in primates, see Dixson (1998), pages 139–45; for the practices of the sooty mangabey, see Gust and Gordon (1991); for sex toys and orangutans, see Rijksen (1978), pages 262–63; for masturbation in chimpanzees, see Temerlin (1976), pages 137–38. For masturbation in red deer, see Darling (1963), pages 160–61. For a theoretical analysis of sperm competition with sneaks and guards, see Parker (1990b). For the basic biology of the plainfin midshipman, see Brantley and Bass (1994); for brain structure, see Bass and Andersen (1991); for gonad mass, see Bass (1992). For details of quarantines and releases of dung beetles in Australia, see Tyndale-Biscoe (1996); for general biology of *Onthophagus binodis*, see Cook (1990); for general biology of *Onthophagus taurus*, see Hunt and Simmons (1998); for differences in big males' investment in sperm production, see Simmons et al. (1999).

## Chapter 6: How to Make Love to a Cannibal

| | |
|---|---|
| European praying mantis | *Mantis religiosa* |
| Garden spider | *Araneus diadematus* |
| Australian redback spider | *Latrodectus hasselti* |
| Sand shark | *Odontaspis taurus* |

### I Like 'Em Headless in Lisbon

For the thrashings of the headless male mantis, see Roeder (1935); for grandmother's footsteps, see Roeder (1935) and Lawrence (1992). For a long (but by no means exhaustive) list of man-eating females, see Elgar (1992). For the gory practices of the midges, see Downes (1978). For the "ate my lover by mistake" hypothesis, see Elgar (1992); for the "only in captivity" hypothesis, see Edmunds (1988); for a list of species observed in and out of the laboratory, see Elgar (1992). For a comparison of the European praying mantis in the wild and in the laboratory, see Lawrence (1992). For spiders that lure males to their doom, see Robinson and Robinson (1980), pages 138–41, and Jackson and Pollard (1997). For female garden spiders prospering as a result of cannibalism, see Elgar and Nash (1988). The notion that cannibalism by females will lead to the evolution of male escape artists is, as far as I know, original to Dr. Tatiana. For spurs in *Tetragnatha extensa* and bondage in *Xysticus cristatus*, see Bristowe (1958), pages 252–56 (for spurs) and pages 115 and 143–45 (for bondage). For the horn of *Argyrodes zonatus*, see Legendre and Lopez (1974) and Lopez and Emerit (1979). For another example of males copulating when they lose their heads, see McDaniel and Horsfall (1957). For erections in throttled human males, see Goldstein (2000).

### Wretched in the Wilderness

For escape behavior in *Paruroctonus mesaensis*, see Polis and Farley (1979); in *Lycosa rabida*, see Rovner (1972). For gastronomic burials in *Argiope aemula*, see Sasaki and Iwahashi (1995); in the bristle worm, see Reish (1957). For a general discussion of what Dr. Tatiana calls platonic cannibalism, see Elgar and Crespi (1992); for intrauterine cannibalism in the sand shark, see Springer (1948). For cannibalism in amoebae, see Waddell (1992); in the paddle crab *Ovalipes catharus*, see Haddon (1995). The shark poem is original to Dr. Tatiana. Sexual cannibalism in hermaphrodites is reported sporadically but has received little attention. For evidence of cannibalism in *Hermissenda crassicornis*, however, see the note on food on page 212 of Bürgin (1965); for their mating technique, see Rutowski (1983). For circumstances where male suicide by cannibalism is expected to evolve, see Buskirk et al. (1984); for a general description of mating in the Australian redback spider, see Forster (1992); for evidence that being eaten is good for males, see Andrade (1996); for evidence that females eat only when hungry, see Andrade (1998).

### Chapter 7: Crimes of Passion

| | |
|---|---|
| Solitary bee | *Anthophora plumipes* |
| Mountain sheep | *Ovis canadensis* |
| Domestic sheep | *Ovis aries* |
| Giant petrel | *Macronectes halli* |

| | |
|---|---|
| Quacking frog | *Crinia georgiana* |
| Wood frog | *Rana sylvatica* |
| Yellow dung fly | *Scatophaga stercoraria* |
| Northern elephant seal | *Mirounga angustirostris* |
| Hawaiian monk seal | *Monachus schauinslandi* |
| Tiger shark | *Galeocerdo cuvier* |
| Water strider | *Gerris odontogaster* |
| Pheasant | *Phasianus colchicus* |
| Pigeon | *Columbia livia* |
| Seaweed fly | *Gluma musgravei* |
| Crabeater seal | *Lobodon carcinophagus* |
| Bison | *Bison bison* |
| Dugong | *Dugong dugon* |
| Pygmy salamander | *Desmognathus wrighti* |
| Southern elephant seal | *Mirounga leonina* |
| Mink | *Mustela vison* |
| Sea otter | *Enhydra lutris* |
| Blue shark | *Prionace glauca* |
| Round stingray | *Urolophus halleri* |
| Sagebrush cricket | *Cyphoderris strepitans* |
| Little brown bat | *Myotis lucifugus* |
| White-fronted bee-eater | *Merops bullockoides* |
| Lesser snow goose | *Chen caerulescens caerulescens* |
| American lobster | *Homarus americanus* |

## A Girl's Never Alone in Oxford

For male stuffing in *Polistes dominulus*, see Starks and Poe (1997). For general biology of the solitary bee *Anthophora plumipes*, see Stone (1995) and Stone et al. (1995). For harassment in *Anthophora plumipes*, see Stone (1995); in mountain sheep, see Geist (1971), page 210. For battery in domestic sheep on the Île Longue, see Réale et al. (1996); for drowning in the quacking frog, see Byrne and Roberts (1999); for drowning in wood frogs, see Banta (1914) and Howard (1980). For dismemberment in solitary bees, see Alcock (1996); in solitary wasps, see Evans et al. (1986). For drowning and dismemberment in yellow dung flies, see Borgia (1981). For battery in northern elephant seals, see Le Boeuf and Mesnick (1990); in the Hawaiian monk seal, see Hiruki, Stirling, et al. (1993) and Hiruki, Gilmartin, et al. (1993). For males having their libidos suppressed, see Nowak (1999), page 869. For bodyguards in yellow dung flies, see Borgia (1980) and Borgia (1981); in northern elephant seals, see Mesnick and Le Boeuf (1991); in water striders, see Arnqvist (1992); in pheasants, see Ridley and Hill (1987); in pigeons, see Lovell-Mansbridge and Birkhead (1998). For male avoidance in *Anthophora plumipes*, see Stone (1995). For revenge in *Philanthus basilaris*, see O'Neill and Evans (1981).

## Mr. Nice Is Mr. Frustrated in Mallacoota Bay

For mating in seaweed flies, see Dunn et al. (2001). For violence, including scarring, in crabeater seals, see Siniff et al. (1979). For violence in bison, see Lott (1981); in dugongs, see Anderson and Birtles (1978). For biting in the pygmy salamander, see Houck (1980) and a personal communication from Nancy Reagan cited in Promislow (1987). For accidents in southern elephant seals, see Carrick and Ingham (1962); in mink, see Enders (1952); in sea otters, see Foott (1970), Riedman and Estes (1990), and Staedler and Riedman (1993). For the suggestion that violence can be beneficial to males by deterring females from remating, see Johnstone and Keller (2000). For a general discussion of reproduction in sharks, see Wourms (1977); for biting in blue sharks, see Stevens (1974) and Pratt (1979); for thick skin in blue sharks, see Pratt (1979); for thick skin and biting in round stingrays, see Nordell (1994); for biting in *Falcatus falcatus*, see Lund (1990). For penis fencing in *Pseudoceros bifurcus*, see Michiels and Newman (1998).

## Don't Know Much about Anatomy in the Rockies

For rape in sagebrush crickets, see Sakaluk et al. (1995); for female feeding behavior, see Morris et al. (1989), Eggert and Sakaluk (1994), and Sakaluk et al. (1995). With respect to rape, scorpionfly species have more or less the same general biology; see Thornhill (1980). My account, however, particularly concerns *Panorpa latipennis*. For the use of the notal organ during rape in this species, see Thornhill (1980); for the burgling of spiderwebs, see Thornhill (1975). For reports of rape in lobsters, see Waddy and Aiken (1991); in fish, see Farr (1980); in turtles, see Berry and Shine (1980); in bats, see Pearson et al. (1952) and Thomas et al. (1979); in birds, see McKinney et al. (1983); in primates, see Mitani (1985) and Smuts and Smuts (1993). For rape in little brown bats, see Thomas et al. (1979). For rape in white-fronted bee-eaters, see Emlen and Wrege (1986); for their general biology, see Emlen and Wrege (1994). For rape in lesser snow geese, see Mineau and Cooke (1979) and Dunn et al. (1999). For rates of conception during rape in white-fronted bee-eaters, see Emlen and Wrege (1986) and Wrege and Emlen (1987); in lesser snow geese, see Dunn et al. (1999). For escape behavior in American lobsters, see Waddy and Aiken (1991); in scorpionflies, see Thornhill (1980); in white-fronted bee-eaters, see Emlen and Wrege (1986). For an example of male birds withdrawing help if they suspect infidelity, see Dixon et al. (1994); for female scorpionflies suffering lower predation rates, see Thornhill (1980). For evidence that females benefit from choosing their own mate, see Partridge (1980) (fruit flies) and Simmons (1987) (field crickets).

## Chapter 8: Hell Hath No Fury

| | |
|---|---|
| Moorhen | *Gallinula chloropus* |
| Seed-harvester ant | *Veromessor pergandei* |

| | |
|---|---|
| Smooth newt | *Triturus vulgaris vulgaris* |
| Darwin frog | *Rhinoderma darwinii* |
| Japanese cardinal fish | *Apogon doederleini* |
| Burying beetle | *Nicrophorus defodiens* |
| House sparrow | *Passer domesticus* |
| Great reed warbler | *Acrocephalus arundinaceus* |
| Pied flycatcher | *Ficedula hypoleuca* |
| Starling | *Sturnus vulgaris* |
| Northern harrier | *Circus cyaneus* |
| Blue tit | *Parus caeruleus* |

## Bring Back the Ladies in Norfolk

For lethal fighting among female thrips, see Crespi (1992); in seed-harvester ants, see Rissing and Pollock (1987). For shortages of males in the smooth newt, see Waights (1996); for sperm limitation in the smooth newt, see Verrell (1986) and Verrell et al. (1986). For female quarreling in katydids, see Gwynne and Simmons (1990). For *Wolbachia* and severe shortages of males in *Acraea encedon*, see Jiggins et al. (1999). For female-female interference in the Majorcan mid-wife toad, see Bush and Bell (1997). For the biology of the Darwin frog, see Cei (1962), pages 110–15, and Goicoechea et al. (1986). For brood cannibalism in the Japanese cardinal fish, see Okuda and Yanagisawa (1996). For fighting between female moorhens, see Petrie (1983).

## I Hate the Trouble and Strife in Ontario

For general biology of burying beetles, see Milne and Milne (1976); for conflict between males and females in *Nicrophorus defodiens*, see Eggert and Sakaluk (1995). For evidence that female burying beetles sharing a carcass have fewer offspring each, see Trumbo and Fiore (1994); I learned of murderous behavior in *Nicrophorus defodiens* in an e-mail from Anne-Katrin Eggert. For egg smashing in the house sparrow, see Veiga (1990); in the great reed warbler, see Hansson et al. (1997). For evidence that female birds often lose male help if their mate takes an additional partner, see Webster (1991) and Slagsvold and Lifjeld (1994); for a general account of female hostility toward mistresses (in birds), see Slagsvold and Lifjeld (1994). For scolding in pied flycatchers, see Slagsvold et al. (1992). For lovey-dovey starlings, see Eens and Pinxten (1996). For aggression in northern harriers, see Simmons (1988); in blue tits, see Kempenaers (1995). For aggression (including singing, and filling up extra nest holes) in starlings, see Eens and Pinxten (1996) and Sandell and Smith (1997). For male starlings chasing their partner away, see Eens and Pinxten (1996) and Pinxten and Eens (1990). For general biology of *Lamprologus ocellatus*, see Brandtmann et al. (1999). For hermit crabs avoiding shells with holes in them, see Pechenik and Lewis (2000). For hydroids preferring to settle on shells occupied by hermit crabs, see Campbell (1974); for hermit crabs gaining protection from hydroids, see Brooks and Gwaltney (1993); for hermit crabs harvesting anemones, see Branch and Branch (1998), caption to plate 67; for hydroids (and anemones) gaining protection from hermit crabs, see Brooks and Gwaltney (1993). For fish in Lake Tanganyika that use secondhand snail shells, and for size differences and shell stealing in *Lamprologus callipterus*, see Sato (1994). For fighting between female *Lamprologus ocellatus*, see Walter and Trillmich (1994) and Brandtmann et al. (1999); for male intervention, see Walter and Trillmich (1994).

## Chapter 9: Aphrodisiacs, Love Potions, and Other Recipes from Cupid's Kitchen

| | |
|---|---|
| Fruit fly | *Drosophila melanogaster* |
| Australian field cricket | *Teleogryllus commodus* |
| Housefly | *Musca domestica* |
| Red deer | *Cervus elaphus* |
| Rock-boring sea urchin | *Echinometra mathaei* |
| Oblong sea urchin | *Echinometra oblonga* |
| Manatee | *Trichechus manatus* |
| Bonobo | *Pan paniscus* |
| Adélie penguin | *Pygoscelis adeliae* |
| Bottle-nosed dolphin | *Tursiops truncatus* |
| Amazon River dolphin | *Inia geoffrensis* |
| Stump-tailed macaque | *Macaca arctoides* |
| Baboon | *Papio anubis* |
| Razorbill | *Alca torda* |
| Japanese macaque | *Macaca fuscata* |
| Rhesus monkey | *Macaca mulatta* |

### Afraid I've Been Bewitched in Santa Barbara

For a general overview of the effects of seminal fluid, see Mann and Lutwak-Mann (1981) and Chen (1984); for the stimulation of egg production in the Australian field cricket, see Loher et al. (1981); for the composition of the seminal fluid of the housefly, see Andrés and Arnqvist (2001); for its effects, see Riemann et al. (1967). For the composition of fruit fly seminal fluid, see Chapman (2001); for male fruit flies using chemicals to disable the sperm of other males, see Harshman and Prout (1994); for male fruit flies protecting their own sperm, see Chapman (2001); for antiaphrodisiacs in fruit flies, see Mane et al. (1983); for the effects of sex peptide, see Wolfner (1997). For *Helix aspersa* and love darts, see Koene and Chase (1998) and Rogers and Chase (2001). For the aphrodisiac effects of roaring in red deer, see McComb (1987); for conception date and calf survival, see McComb (1987) and Clutton-Brock et al. (1988). For female houseflies resisting local males but not strangers, see Andrés and Arnqvist (2001). For the effects of forced monogamy on fruit flies, see Pitnick et al. (2001). Information on shrimp that live in glass sponges is sparse, but see Berggren (1993) and Saito and Konishi (1999). For the evolution of supermales in fruit flies, see Rice (1996).

## Desperate to Be à la Mode in Hawaii

For bindin-egg affinities in the rock-boring sea urchin, see Palumbi (1999); for genetic differences between sea urchin species, see Palumbi and Metz (1991). For rapid evolution of reproductive proteins in sea urchins, see Metz and Palumbi (1996); in mammals, see Swanson et al. (2001); in fruit flies, see Aguadé (1999); in abalone, see Metz et al. (1998). For the notion that sexual conflict can drive the origin of species, see Rice and Hostert (1993); for speciation and rapid evolution of reproductive proteins, see Swanson and Vacquier (2002); for the association between the formation of new insect species and female remating rate, see Arnqvist et al. (2000). For the evolution of the abalone VERL and lysin, see Metz et al. (1998), Swanson and Vacquier (1997), Swanson and Vacquier (1998), Yang et al. (2000), and Swanson et al. (2001). For the evolution of genes containing short repeated units and their role in human disease, see Mitas (1997); for concerted evolution—the evolution of genes containing large repeated units—see Elder and Turner (1995).

## Don't Want No Homo in the Florida Keys

For homosexual behavior in bonobos, see de Waal (1989), pages 201–04; in penguins, see Davis et al. (1998); in bottle-nosed dolphins, see McBride and Hebb (1948), Brown and Norris (1956), and Tavolga (1966); in Amazon River dolphins, see Pilleri et al. (1980), Sylvestre (1985), and Best and da Silva (1989); in manatees, see Hartman (1979). For a general source on homosexual behavior in animals, see the appendix to Bagemihl (1999). For orgasms in homosexual interactions in female stump-tailed macaques, see Goldfoot et al. (1980); in heterosexual interactions, see Slob et al. (1986). For homosexual behavior and cooperation in baboons, see Smuts and Watanabe (1990). For homosexual mounting in razorbills, see Wagner (1996). For homosexual octopuses, see Lutz and Voight (1994); for lesbian gulls, see Kovacs and Ryder (1983), Hunt et al. (1984), and Conover and Hunt (1984). For reports of genes involved in human homosexuality, see Hamer et al. (1993) and Hu et al. (1995); for a failure to replicate these results, see Rice et al. (1999). For homosexual behavior in fruit flies, see Hall (1994), Ryner et al. (1996), and Yamamoto et al. (1997). For flies that won't have sex in the dark, see Sharma (1977). For competition for female Japanese macaques, see Vasey (1998); for anal sex in rhesus monkeys, see Erwin and Maple (1976). For a discussion of the difficulties of measuring the prevalence of homosexuality in humans, see LeVay (1996), chapter 2. For general discussions of the evolution of kin-directed altruism, see Hamilton (1996); for the notion that this could explain homosexuality, see Wilson (1975), pages 343–44. For social structure and reproductive suppression in termites, see Wilson (1971), chapter 10; in wolves, see Nowak (1999), page 667; in naked mole rats,

see Faulkes and Bennett (2001); in shrimp, see Duffy (1996). I believe that
Hutchinson (1959) was the first to posit heterozygote advantage as a possible
explanation for homosexuality. For heterozygote advantage and resistance to
malaria, see any genetics textbook. For a theoretical treatment of genes beneficial in one sex spreading despite being detrimental in the other, see Rice (1984);
for evidence, see Chippindale et al. (2001). The notion that this could account
for the evolution of homosexuality is, as far as I know, original to Dr. Tatiana.

## Chapter 10: Till Death Do Us Part

| | |
|---|---|
| Black vulture | *Coragyps atratus* |
| Gibbon | *Hylobates lar* |
| Jackdaw | *Corvus monedula* |
| Chinstrap penguin | *Pygoscelis antarctica* |
| Long-eared owl | *Asio otus* |
| Kirk's dik-dik | *Madoqua kirkii* |
| California mouse | *Peromyscus californicus* |
| Termite | *Reticulitermes flavipes* |
| African hawk eagle | *Hieraaetus spilogaster* |
| Fat-tailed dwarf lemur | *Cheirogaleus medius* |
| Djungarian hamster | *Phodopus campbelli* |
| Siberian hamster | *Phodopus sungorus* |
| Banded shrimp | *Stenopus hispidus* |
| Wreathed hornbill | *Aceros undulatus* |
| Bewick's swan | *Cygnus columbianus bewickii* |
| Prairie vole | *Microtus ochrogaster* |
| Montane vole | *Microtus montanus* |
| Indian crested porcupine | *Hystrix indica* |
| Southern elephant seal | *Mirounga leonina* |
| Gorilla | *Gorilla gorilla* |
| Chimpanzee | *Pan troglodytes* |

### *Crusading for Family Values in Louisiana*

For early beliefs about monogamy in birds, see Lack (1968), especially chapter 14.
For adultery in gibbons, see Sommer and Reichard (2000). For fidelity in the black
vulture, see Decker et al. (1993); in the jackdaw, see Henderson et al. (2000); in the
chinstrap penguin, see Moreno et al. (2000); in the long-eared owl, see Marks et
al. (1999); in Kirk's dik-dik, see Brotherton et al. (1997); in the California mouse,
see Ribble (1991). Termite monogamy refers to outbreeding only, see Bulmer et
al. (2001). For the notion that a male should attempt a mixed strategy—helping
one female rear offspring while siring, by other females, children that he will not
help rear, see Trivers (1972). For the lack of paternal care in Kirk's dik-dik, see
Brotherton and Rhodes (1996). For female fat-tailed dwarf lemurs' inability to
raise young as single mothers, see Fietz (1999); for their infidelity, see Fietz et al.

(2000). For a discussion of the relationship between biparental care and monogamy, see Komers and Brotherton (1997). For the importance of spatial distribution of females in the evolution of monogamy, see Komers and Brotherton (1997). For danger maintaining monogamy in *Lysiosquilla sulcata*, see Caldwell (1991). For general biology of the Djungarian hamster, see Nowak (1999), page 1419; for the male acting as midwife, see Jones and Wynne-Edwards (2000); for contrasts with the Siberian hamster, see Wynne-Edwards (1995). For pair formation and violence in banded shrimp, see Johnson (1969); for feeding and molting, see Hoover (1998), page 219. For breeding biology of hornbills, see Kemp (1995), chapter 5. For breeding biology of jackdaws, see Henderson et al. (2000); for pair duration and reproductive success in Bewick's swans, see Rees et al. (1996). For prudish behavior of black vultures, see Decker et al. (1993).

## So Bummed Out in Berkeley

For self-restraint among male California mice, see Gubernick and Nordby (1993); for monogamy in California mice, see Ribble (1991). For excellent overviews of the hormonal mechanisms of monogamy, see Young et al. (1998) and Insel and Young (2001). For the role of vasopressin in prairie vole monogamy, see Winslow et al. (1993); for transgenic mice and vasopressin, see Young et al. (1999); for behavioral contrasts with montane voles, see Shapiro and Dewsbury (1990); for hormonal contrasts, see Young et al. (1998). For frequent sex in the Indian crested porcupine, see Sever and Mendelssohn (1988); in the wood roach, see Nalepa (1988). For nonidentical twins having different fathers, see Phelan et al. (1982). For size differences between male and female southern elephant seals, see Nowak (1999), page 880; between male and female gorillas, see Nowak (1999), page 620. For differences in testicle size between humans and other apes, see Short (1979). For an excellent discussion of the fiction of high rates of infidelity as measured by paternity testing in humans, see Macintyre and Sooman (1991); for studies showing low rates of infidelity, see Ashton (1980) and Sasse et al. (1994); for the rate of 11.8 percent, see Cerda-Flores et al. (1999); for infidelity among Sykeses, see Sykes and Irven (2000). For a genetic predisposition toward promiscuity in crickets, see Solymar and Cade (1990); in fruit flies, see Pyle and Gromko (1981).

## Chapter 11: The Fornications of Kings

| | |
|---|---|
| Lesser mealworm beetle | *Alphitobius diaperinus* |
| White-lipped land snail | *Triodopsis albolabris* |
| True armyworm moth | *Pseudaletia unipuncta* |
| Woodlouse | *Armadillidium vulgare* |
| Wood lemming | *Myopus schisticolor* |
| House sparrow | *Passer domesticus* |
| Great land crab | *Cardisoma guanhumi* |
| Pink salmon | *Oncorhynchus gorbuscha* |

| Soapberry bug | *Jadera haematoloma* |
|---|---|
| Soapberry tree | *Sapindus saponaria* |
| Round-podded golden rain tree | *Koelreuteria paniculata* |

## Aghast in Arkansas

For the natural history of *Acarophenax mahunkai*, see Steinkraus and Cross (1993). Many thanks to Don Steinkraus for confirming that the mites' ability to detect the sex of beetles is probable but unknown. For the effects of close incest in humans, see Seemanová (1971). For a general account of inbreeding depression and recessive genes, see any population genetics textbook; for a more technical account of the subject (including discussions of alternative, but less widely accepted explanations for inbreeding depression that are not mentioned by Dr. Tatiana), see Charlesworth and Charlesworth (1987). For marriages on kibbutzim, see Shepher (1971). For self-copulation in *Dendrobaena rubida*, see André and Davant (1972). For the evolution of selfing rates in hermaphrodites, see Jarne and Charlesworth (1993); for emergency selfing in the white-lipped land snail, see McCracken and Brussard (1980). For Hawaiian incest, see Malo (1903), chapter 18; for Egyptian incest, see Tyldesley (1994), pages 198–99; for a general discussion of royal incest (including the notion that it is a result of stratification) and for incest among the Incas, see van den Berghe and Mesher (1980). For the independent origins of haplodiploidy and paternal genome elimination, see Mable and Otto (1998). For haplodiploidy and inbreeding in pinworms, see Adamson (1989); in insects and mites (and for discussions of paternal genome elimination), see Wrensch and Ebbert (1993). For incest in the button beetle, see Hamilton (1993), page 430. For the life and times of *Scleroderma immigrans*, see Wheeler (1928), pages 62–64. The advantage of having haploid males when making a switch to incest has been remarked on by many authors; for a formal treatment, see Werren (1993).

## Piqued in Darien

For a marvelous account of the biology of the moth ear mite, see Treat (1975), chapter 7; for the sex ratio in moth ear mites, as well as proof of their paternal genome elimination, see Treat (1965). The moth nursery rhyme is original to Dr. Tatiana. For mites on the antennae and on the feet of army ants, for hummingbird nostril mites, and for fruit bat eyeball mites, see Walter and Proctor (1999), pages 200–201 (Dave Walter tells me that some species of army ant are likely to have both types of mite); for human follicle mites, see Nutting (1976); for quill mites, see Kethley (1971). For Fisher's argument on sex ratios, see Fisher (1999), pages 141–43. For the effect of war on the human sex ratio, see Graffelman and Hoekstra (2000). For feminization of genetic males in woodlice, see Bull (1983), page 200; for sex ratio skew in wood lemmings, see Nowak (1999), pages 1482–83; for their maverick chromosomes, see Bull (1983), pages 79–80. For sex ratios in quill mites, see Kethley (1971). For sex ratios in

*Acarophenax mahunkai*, see Steinkraus and Cross (1993). For the evolution of sex ratios under inbreeding, see Hamilton (1996), chapter 4. For selfing rates and allocation to male tissue in *Utterbackia imbecillis*, see Johnston et al. (1998); for smaller flowers in selfing plants, see Charnov (1982), pages 261–68; for runty males in incestuous animal species, see Hamilton (1993). For sex ratio adjustment in *Nasonia vitripennis*, see Werren (1980). For control of sex ratio under paternal genome elimination in *Typhlodromus occidentalis*, see Nagelkerke and Sabelis (1998).

## Gagging for It in Florida

For mangrove fish sharing burrows with land crabs and for their ability to survive out of water, see Taylor (1990). For outbreeding depression in pink salmon, see Gharrett et al. (1999). My discussion of the soapberry bug is a simplification of Carroll and Boyd (1992). For male frequencies in *Rivulus marmoratus*, see Taylor et al. (2001); for effects of outcrossing, see Taylor (2001).

## Chapter 12: Eve's Testicle

| | |
|---|---|
| Black hamlet fish | *Hypoplectrus nigricans* |
| Green spoon worm | *Bonellia viridis* |
| Spotted hyena | *Crocuta crocuta* |
| Brown hyena | *Hyaena brunnea* |
| Striped hyena | *Hyaena hyaena* |
| Aardwolf | *Proteles cristatus* |
| Wildebeest | *Connochaetes taurinus* |
| Thomson's gazelle | *Gazella thomsonii* |
| Paper nautilus | *Argonauta argo* |
| Green poison arrow frog | *Dendrobates auratus* |
| Spraying characid | *Copeina arnoldi* |
| Dayak fruit bat | *Dyacopterus spadiceus* |

## Looking for a Baker's Dozen in the Forests of Romania

For an excellent description of the life cycle of *Physarum polycephalum*, see Bailey (1997); for the system of sexes, see Kawano et al. (1987) and Bailey (1997); for *matA* and the removal of the mitochondria after fusion, see Meland et al. (1991). My definition of sexes follows standard practice. For predictions that if finding a mate is difficult, isogamous organisms should have a large number of sexes, see Iwasa and Sasaki (1987); for the notion that having two sexes is deeply strange, see Hurst (1996); for the efficiency of inbreeding prevention and outbreeding promotion in organisms with multiple mating types, see Raper (1966). For problems evolving from zero to two sexes, see Hoekstra (1987) and Hutson and Law (1993). For the notion that sexes evolved to prevent conflict between cytoplasmic elements, see Eberhard (1980), Cosmides and Tooby (1981), Hurst and Hamilton (1992), Hurst (1996). For mitochondria and chloroplast transmission in *Chlamydomonas reinhardtii*, see Gillham (1994). For large numbers of sexes in mushrooms and ciliates, see references in Hurst and Hamilton (1992); for twenty thousand sexes in *Schizophyllum commune*, see Kothe (1996); for difficulties in regulating the transmission of genetic elements when there are more than two sexes, see Hurst (1996); for badly behaved slime mold mitochondria, see Kawano et al. (1991).

## Ready to Litigate in Tallahassee

The problem of evolving males and females from isogamy has been discussed extensively. For a recent review, see Randerson and Hurst (2001). For *Chlamydomonas moewusii* producing small cells, see Weise et al. (1979). For the advantage of being a male, see Parker et al. (1972)—for problems with the theory, see Randerson and Hurst (2001). In arguing that bigness is an advantage because it makes you easier to find, I follow Dusenbery (2000). For the notion that males and females evolved to control cytoplasmic elements, see Hastings (1992); for problems with this idea, see Randerson and Hurst (2001).

## Group Sexists in Santa Catalina

For orgies in *Aplysia californica*, see Pennings (1991); for spawning behavior of the black hamlet fish, see Fischer (1980). For fusion in *Diplozoon gracile*, see Justine et al. (1985). For aerial sex in *Limax maximus*, see Langlois (1965); in *Limax redii*, see Baur (1998). For a superb general treatment of when to be a hermaphrodite, see Charnov (1982); for low density and hermaphrodites, see Ghiselin (1969); for the prediction that wind pollination leads to separate sexes, see Charnov et al. (1976). For wind-pollinated plants tending to having separate sexes and pollinator-pollinated plants tending toward hermaphroditism, see Renner and Ricklefs (1995). For hermaphroditism in fish, see Charnov (1982), table 12.1; for deep-sea comb jellies with separate sexes, see Harbison and Miller (1986); for patterns of hermaphroditism in bivalves and barnacles, see Charnov (1982), page 239; for the absence of hermaphrodites from many groups, see Ghiselin (1974), chapter 4. Little has been written about the transition between separate sexes and hermaphroditism. For some of the genetic considerations, see White (1973). For males, females, and hermaphrodites in the cactus *Pachycereus pringlei*, see Fleming et al. (1994).

## Too Much Heavy Breathing near Malta

For dwarf males and sex determination in the green spoon worm, see Jaccarini et al. (1983). Many thanks to René Hessling for calculating the size difference between males and females. For a general overview of sex determination, see Bull (1983); for sex chromosomes, see particularly pages 16–20; for patterns of sex determination in reptiles, see pages 115–22. For multiple independent origins of males hatching from unfertilized eggs, see Mable and Otto (1998), table 1; for sex determination by fungal infection in *Stictococcus sjoestedti*, see Bacci

(1965), pages 154–55. For *Capitella* becoming a hermaphrodite at low density, see Holbrook and Grassle (1984). For sex change in the slipper limpet, see Hoagland (1978); in *Ophryotrocha puerilis*, see Berglund (1986). For a general account of when to change sex, see Charnov (1982); for the size advantage model of sex change, see Ghiselin (1969). For examples of dwarf males in other species and for the notion that dwarf males are favored when females are both sedentary and few and far between, see Ghiselin (1974), chapter 7; for dwarf males in anglerfish, and for general anglerfish biology, see Bertelsen (1951); for the claim that anglerfish males have the largest noses in proportion to their bodies, see Andersson (1994), page 257.

## Don't Wanna Be Butch in Botswana

For the hunting and dining habits of spotted hyenas and for lions scavenging from hyenas, see Kruuk (1972), especially chapter 5; for comparisons with other members of the hyena family, see Estes (1991), chapter 20, and Nowak (1999), pages 786–93. For the social structure of spotted hyenas, see Kruuk (1972), chapter 6, and Frank (1997). For the belief that the spotted hyena was a hermaphrodite and for the structure of the female's genitalia, see Kruuk (1972), page 210; for copulation in the spotted hyena, see Frank (1997); for the structure of the birth canal and for estimated death rates during birth, see Frank et al. (1995). For discussion of the idea that the female's phallus has evolved because of its use in greeting ceremonies, see Kruuk (1972), chapter 6, and Frank (1997); for the theory that the phallus is an antirape device, see East et al. (1993); for a vigorous debunking of the antirape theory, see Frank (1997). For aggressiveness in female mice and its relationship to position in the womb, see vom Saal (1989); for androgen exposure and genital abnormalities, see Frank (1997). For siblicide in spotted hyenas, see Frank et al. (1991); for greater reproductive success of dominant females, see Frank (1997). For the blocking of androgen exposure in the womb and its lack of effect on female genitalia in hyenas, see Drea et al. (1998). For spiders charging their pedipalps, see Bristowe (1958), pages 65–67; for genitalia in the seahorse, see Eberhard (1985), page 68. For the genitalia of *Sapha amicorum*, see Marcus (1959). For the penis of the paper nautilus, see Müller (1853) and Young (1959). For the pouch in the leech *Marsupiobdella africana*, see van der Lande and Tinsley (1976); for hunting in *Helobdella striata*, see Kutschera and Wirtz (1986). For parental care in the green poison arrow frog, see Wells (1978) and Summers (1989). For the spraying characid laying and tending eggs, see Krekorian (1976). For male Dayak fruit bats producing milk, see Francis et al. (1994). The gender bender poem is original to Dr. Tatiana.

## Chapter 13: Wholly Virgin

For general biology of bdelloid rotifers, see Donner (1966) and Ricci (1987). David Mark Welch kindly told me that the age of the bdelloid rotifers, based on molecular evidence, is eighty-five million years. The problem of the evolution of sex has been discussed many times: see, for example, Maynard Smith (1978), Bell (1982), and Kondrashov (1993). For the particular problem posed by the bdelloid rotifers, see Maynard Smith (1986) and Judson and Normark (1996). Maynard Smith (1986) dubbed the bdelloid rotifers an "evolutionary scandal." My description of the cost of sex is drawn from Maynard Smith (1978), page 3. My account of bacterial sex is based on Levin (1988), Maynard Smith et al. (1993), and Davies (1994); for viral sex, see Chao (1992). An account of meiosis can be found in any basic genetics textbook.

For other putative ancient asexuals, see Judson and Normark (1996), although for the age of the darwinulid ostracods, see Butlin et al. (1999). For the discrediting of the chaetonotid gastrotrichs, see Weiss and Levy (1979); of the aphids, see Normark (1999). Ben Normark kindly told me of the discovery of males. Hurst et al. (1992) and Little and Hebert (1996) cast doubt on the existence of ancient asexuals. For the pattern of molecular evolution that you expect to observe in an ancient asexual lineage, see Judson and Normark (1996) and Birky (1996). For proof that the bdelloid rotifers have the expected pattern and have therefore been without sex for millions of years, see Mark Welch and Meselson (2000).

The notion that asexuals cannot evolve is extremely common. For the number and general distribution of bdelloid species, see Ricci (1987); for evidence that bdelloids occur in the Antarctic, see Everitt (1981); for the lives of seisonid rotifers, see Ricci (1992). In arguing that most mutations are neutral, Miss *Philodina* is adopting the position of Kimura (1983).

Muller (1964) proposed the ratchet. It has been commented on extensively; for how asexuals can evade the ratchet, see Judson and Normark (1996). The hatchet mechanism was proposed by Kondrashov (1984) and Kondrashov (1988). For a general account of *Atta colombica* and the fungus, see Wilson (1971), pages 41–48; for evidence that the fungus is an ancient asexual, see Chapela et al. (1994), especially note 19; for all details of *Escovopsis*, see Currie et al. (1999). The idea that parasites are important in maintaining sex and variation was suggested by Haldane (1949), Bremermann (1980), Hamilton (1980), and Tooby (1982); Bell (1982) christened this theory the Red Queen. See Carroll (1871) for the Red Queen quotation. For clonal reproduction in armadillos, see Loughry et al. (1998); for the armadillo's penis, see Wassersug (1997). Ladle et al. (1993) suggested ancient asexuals can escape the Red Queen if they disperse without parasites. For the survival of bdelloid rotifers after anhydrobiosis, see Ricci (1987).

## Postscript

For the idea that sex arose through cannibalism, see Margulis and Sagan (1986); for the notion that it arose through the action of genetic elements trying to promote their own spread, see Hickey and Rose (1988); for the notion that it arose to enable DNA repair, see Bernstein et al. (1988).

# BIBLIOGRAPHY

Abele, L. G., and S. Gilchrist, 1977. Homosexual rape and sexual selection in Acanthocephalan worms. *Science* 197: 81–83.

Adamson, M. L., 1989. Evolutionary biology of the oxyurida (Nematoda): Biofacies of a haplodiploid taxon. *Advances in Parasitology* 28: 175–228.

Afzelius, B. A., B. Baccetti, and R. Dallai, 1976. The giant spermatozoon of *Notonecta. Journal of Submicroscopic Cytology* 8: 149–61.

Afzelius, B. A., and R. Dallai, 1983. The paired spermatozoa of the marine snail, *Turritella communis* Lamarck (Mollusca, Mesogastropoda). *Journal of Ultrastructure Research* 85: 311–19.

Aguadé, M., 1999. Positive selection drives the evolution of the *Acp29AB* accessory gland protein in *Drosophila. Genetics* 152: 543–51.

Ahlgren, M., 1975. Sperm transport to and survival in the human fallopian tube. *Gynecologic Investigation* 6: 206–14.

Alatalo, R. V., J. Höglund, A. Lundberg, and W. J. Sutherland, 1992. Evolution of black grouse leks: Female preferences benefit males in larger leks. *Behavioral Ecology* 3: 53–59.

Alcock, J., 1996. The relation between male body size, fighting, and mating success in Dawson's burrowing bee, *Amegilla dawsoni* (Apidae, Apinae, Anthophorini). *Journal of Zoology* 239: 663–74.

Alcock, J., G. C. Eickwort, and K. R. Eickwort, 1977. The reproductive behavior of *Anthidium maculosum* (Hymenoptera: Megachilidae) and the evolutionary significance of multiple copulations by females. *Behavioral Ecology and Sociobiology* 2: 385–96.

Aldrich, J. M., and L. A. Turley, 1899. A balloon-making fly. *American Naturalist* 33: 809–12.

Anderson, P. K., and A. Birtles, 1978. Behaviour and ecology of the dugong, *Dugong dugon* (Sirenia): Observations in Shoalwater and Cleveland Bays, Queensland. *Australian Wildlife Research* 5: 1–23.

Andersson, M., 1994. *Sexual Selection*. Princeton University Press.

Andrade, M. C. B., 1996. Sexual selection for male sacrifice in the Australian redback spider. *Science* 271: 70–72.

———, 1998. Female hunger can explain variation in cannibalistic behavior despite male sacrifice in redback spiders. *Behavioral Ecology* 9: 33–42.

André, F., and N. Davant, 1972. L'autofécondation chez les lombriciens. Observation d'un cas d'autoinsémination chez *Dendrobaena rubida* f. *subrubicunda Eisen*. *Bulletin de la Société Zoologique de France* 97: 725–28.

Andrés, J. A., and G. Arnqvist, 2001. Genetic divergence of the seminal signal-receptor system in houseflies: The footprints of sexually antagonistic coevolution? *Proceedings of the Royal Society of London, B* 268: 399–405.

Anstett, M. C., M. Hossaert-McKey, and F. Kjellberg, 1997. Figs and fig pollinators: Evolutionary conflicts in a coevolved mutualism. *Trends in Ecology and Evolution* 12: 94–99.

Arnold, S. J., 1976. Sexual behavior, sexual interference, and sexual defense in the salamanders *Ambystoma maculatum, Ambystoma tigrinum,* and *Plethodon jordani*. *Zeitschrift für Tierpsychologie* 42: 247–300.

Arnqvist, G., 1992. Pre-copulatory fighting in a water strider: Inter-sexual conflict or mate assessment? *Animal Behaviour* 43: 559–67.

———, 1998. Comparative evidence for the evolution of genitalia by sexual selection. *Nature* 393: 784–86.

Arnqvist, G., M. Edvardsson, U. Friberg, and T. Nilsson, 2000. Sexual conflict promotes speciation in insects. *Proceedings of the National Academy of Sciences, USA* 97: 10460–64.

Ashton, G. C., 1980. Mismatches in genetic markers in a large family study. *American Journal of Human Genetics* 32: 601–13.

Atkinson, W. D., 1979. A comparison of the reproductive strategies of domestic species of *Drosophila*. *Journal of Animal Ecology* 48: 53–64.

Austad, S. N., and R. Thornhill, 1986. Female reproductive variation in a nuptial-feeding spider, *Pisaura mirabilis*. *Bulletin/British Arachnological Society* 7: 48–52.

Bacci, G., 1965. *Sex Determination*. Pergamon Press.

Bagemihl, B., 1999. *Biological Exuberance*. St. Martin's Press.

Bailey, J., 1997. Building a plasmodium: Development in the acellular slime mould *Physarum polycephalum*. *BioEssays* 19: 985–92.

Banta, A. M., 1914. Sex recognition and the mating behavior of the wood frog, *Rana sylvatica*. *Biological Bulletin* 26: 171–83.

Barker, D. M., 1992. Evolution of sperm shortage in a selfing hermaphrodite. *Evolution* 46: 1951–55.

———, 1994. Copulatory plugs and paternity assurance in the nematode *Caenorhabditis elegans*. *Animal Behaviour* 48: 147–56.

Barratt, C. L. R., A. E. Bolton, and I. D. Cooke, 1990. Functional significance of white blood cells in the male and female reproductive tract. *Human Reproduction* 5: 639–48.

Bass, A., 1992. Dimorphic male brains and alternative reproductive tactics in a vocalizing fish. *Trends in Neuroscience* 15: 139–45.

Bass, A., and K. Andersen, 1991. Inter- and intrasexual dimorphisms in the vocal control system of a teleost fish: Motor axon number and size. *Brain, Behavior and Evolution* 37: 204–14.

Bateman, A. J., 1948. Intra-sexual selection in *Drosophila*. *Heredity* 2: 349–68.

Baur, B., 1998. Sperm competition in molluscs. In *Sperm Competition and Sexual Selection*, ed. T. R. Birkhead and A. P. Møller. Academic Press.

Baylis, J. R., 1981. The evolution of parental care in fishes, with reference to Darwin's rule of male sexual selection. *Environmental Biology of Fishes* 6: 223–51.

Beatty, R. A., 1960. Fertility of mixed semen from different rabbits. *Journal of Reproduction and Fertility* 1: 52–60.

Beauchamp, G. K., K. Yamazaki, and E. A. Boyse, 1985. The chemosensory recognition of genetic individuality. *Scientific American* 253 (July): 66–72.

Beer, A. E., J. F. Quebbeman, and J. W. T. Ayers, 1982. The immunobiology of abortion. In *Immunological Factors in Human Reproduction*, ed. S. Shulman, F. Dondero, and M. Nicotra. Academic Press.

Bell, G., 1982. *The Masterpiece of Nature: The Evolution and Genetics of Sexuality*. University of California Press.

Bell, P. D., 1979. Acoustic attraction of herons by crickets. *New York Entomological Society* 87: 126–27.

Berggren, M., 1993. *Spongiocaris hexactinellicola*, a new species of stenopodidean shrimp (Decapoda: Stenopodidae) associated with hexactinellid sponges from Tartar Bank, Bahamas. *Journal of Crustacean Biology* 13: 784–92.

Berglund, A., 1986. Sex change by a polychaete: Effects of social and reproductive costs. *Ecology* 67: 837–45.

Bernstein, H., F. A. Hopf, and R. E. Michod, 1988. Is meiotic recombination an adaptation for repairing DNA, producing genetic variation, or both? In *The Evolution of Sex: An Examination of Current Ideas*, ed. R. E. Michod and B. R. Levin. Sinauer.

Berry, J. F., and R. Shine, 1980. Sexual size dimorphism and sexual selection in turtles (Order Testudines). *Oecologia* 44: 185–91.

Bertelsen, E., 1951. *The Ceratioid Fishes: Ontogeny, Taxonomy, Distribution and Biology*. DANA-Report 39. Carlsberg Foundation.

Best, R. C., and V. M. F. da Silva, 1989. Amazon river dolphin, Boto *Inia geoffrensis* (de Blainville, 1817). In *Handbook of Marine Mammals*. Vol. 4: *River Dolphins and the Larger Toothed Whales*, ed. S. H. Ridgway and R. Harrison. Academic Press.

Bierzychudek, P., 1981. Pollinator limitation of plant reproductive effort. *American Naturalist* 117: 838–40.

————, 1987. Pollinators increase the cost of sex by avoiding female flowers. *Ecology* 68: 444–47.

Birkhead, T. R., 1998. Cryptic female choice: Criteria for establishing female sperm choice. *Evolution* 52: 1212–18.

Birkhead, T. R., F. Fletcher, E. J. Pellatt, and A. Staples, 1995. Ejaculate quality and the success of extra-pair copulations in the zebra finch. *Nature* 377: 422–23.

Birkhead, T. R., A. P. Møller, and W. J. Sutherland, 1993. Why do females make it so difficult for males to fertilize their eggs? *Journal of Theoretical Biology* 161: 51–60.

Birky, C. W., Jr., 1996. Heterozygosity, heteromorphy, and phylogenetic trees in asexual eukaryotes. *Genetics* 144: 427–37.

Blumer, L. S., 1979. Male parental care in the bony fishes. *Quarterly Review of Biology* 54: 149–61.

Boarder, A., 1968. Coldwater queries. *Aquarist and Pondkeeper* 33: 430–31.

Borgia, G., 1980. Sexual competition in *Scatophaga stercoraria*: Size- and density-related changes in male ability to capture females. *Behaviour* 75: 185–206.

————, 1981. Mate selection in the fly *Scatophaga stercoraria*: Female choice in a male-controlled system. *Animal Behaviour* 29: 71–80.

————, 1985a. Bower destruction and sexual competition in the satin bower-bird (*Ptilonorhynchus violaceus*). *Behavioral Ecology and Sociobiology* 18: 91–100.

————, 1985b. Bower quality, number of decorations, and mating success of male satin bowerbirds (*Ptilonorhynchus violaceus*): An experimental analysis. *Animal Behaviour* 33: 266–71.

Borgia, G., and M. A. Gore, 1986. Feather stealing in the satin bowerbird (*Ptilonorhynchus violaceus*): Male competition and the quality of display. *Animal Behaviour* 34: 727–38.

Bourgeron, T., 2000. Mitochondrial function and male infertility. *Results and Problems in Cell Differentiation* 28: 187–210.

Bradbury, J. W., 1977. Lek mating behavior in the hammer-headed bat. *Zeitschrift für Tierpsychologie* 45: 225–55.

Branch, G., and M. Branch, 1998. *The Living Shores of Southern Africa*. Struik.

Brandtmann, G., M. Scandura, and F. Trillmich, 1999. Female-female conflict in the harem of a snail cichlid (*Lamprologus ocellatus*): Behavioural interactions and fitness consequences. *Behaviour* 136: 1123–44.

Brantley, R. K., and A. H. Bass, 1994. Alternative male spawning tactics and acoustic signals in the plainfin midshipman fish *Porichthys notatus* Girard (Teleostei, Batrachoididae). *Ethology* 96: 213–32.

Bremermann, H. J., 1980. Sex and polymorphism as strategies in host-pathogen interactions. *Journal of Theoretical Biology* 87: 671–702.

Bressac, C., A. Fleury, and D. Lachaise, 1994. Another way of being anisoga-mous in *Drosophila* subgenus species: Giant sperm, one-to-one gamete ratio, and high zygote provisioning. *Proceedings of the National Academy of Sciences, USA* 91: 10399–402.

Bristowe, W. S., 1958. *The World of Spiders.* Collins.

Brockmann, H. J., and D. Penn, 1992. Male mating tactics in the horseshoe crab, *Limulus polyphemus. Animal Behaviour* 44: 653–65.

Brooker, M. G., I. Rowley, M. Adams, and P. R. Baverstock, 1990. Promiscuity: An inbreeding avoidance mechanism in a socially monogamous species? *Behavorial Ecology and Sociobiology* 26: 191–99.

Brooks, W. R., and C. L. Gwaltney, 1993. Protection of symbiotic cnidarians by their hermit crab hosts: Evidence for mutualism. *Symbiosis* 15: 1–13.

Brotherton, P. N. M., J. M. Pemberton, P. E. Komers, and G. Malarky, 1997. Genetic and behavioural evidence of monogamy in a mammal, Kirk's dik-dik (*Madoqua kirkii*). *Proceedings of the Royal Society of London, B* 264: 675–81.

Brotherton, P. N. M., and A. Rhodes, 1996. Monogamy without biparental care in a dwarf antelope. *Proceedings of the Royal Society of London, B* 263: 23–29.

Brown, D. H., and K. S. Norris, 1956. Observations of captive and wild cetaceans. *Journal of Mammalogy* 37: 311–26.

Bull, J. J., 1983. *Evolution of Sex Determining Mechanisms.* Benjamin/Cummings.

Bulmer, M. S., E. S. Adams, and J. F. A. Traniello, 2001. Variation in colony structure in the subterranean termite *Reticulitermes flavipes. Behavioral Ecology and Sociobiology* 49: 236–43.

Burd, M., 1994. Bateman's Principle and plant reproduction: The role of pollen limitation in fruit and seed set. *Botanical Review* 60: 83–139.

Bürgin, U. F., 1965. The color pattern of *Hermissenda crassicornis* (Eschscholtz, 1831) (Gastropoda: Opisthobranchia: Nudibranchia). *Veliger* 7: 205–15.

Burk, T., 1982. Evolutionary significance of predation on sexually signalling males. *Florida Entomologist* 65: 90–104.

Burley, N., G. Krantzberg, and P. Radman, 1982. Influence of colour-banding on the conspecific preferences of zebra finches. *Animal Behaviour* 30: 444–55.

Bush, S. L., and D. J. Bell, 1997. Courtship and female competition in the Majorcan midwife toad, *Alytes muletensis. Ethology* 103: 292–303.

Buskirk, R. E., C. Frohlich, and K. G. Ross, 1984. The natural selection of sexual cannibalism. *American Naturalist* 123: 612–25.

Butchart, S. H. M., N. Seddon, and J. M. M. Ekstrom, 1999. Yelling for sex: Harem males compete for female access in bronze-winged jacanas. *Animal Behaviour* 57: 637–46.

Butlin, R. K., I. Schön, and K. Martens, 1999. Origin, age, and diversity of clones. *Journal of Evolutionary Biology* 12: 1020–22.

Butlin, R. K., C. W. Woodhatch, and G. M. Hewitt, 1987. Male spermatophore investment increases female fecundity in a grasshopper. *Evolution* 41: 221–25.

Bygott, J. D., B. C. R. Bertram, and J. P. Hanby, 1979. Male lions in large coalitions gain reproductive advantages. *Nature* 282: 839–41.

Byrne, P. G., and J. D. Roberts, 1999. Simultaneous mating with multiple males reduces fertilization success in the myobatrachid frog *Crinia georgiana*. *Proceedings of the Royal Society of London, B* 266: 717–21.

Cade, W., 1975. Acoustically orienting parasitoids: Fly phonotaxis to cricket song. *Science* 190: 1312–13.

———, 1979. The evolution of alternative male reproductive strategies in field crickets. In *Sexual Selection and Reproductive Competition in Insects*, ed. M. S. Blum and N. A. Blum. Academic Press.

———, 1981. Alternative male strategies: Genetic differences in crickets. *Science* 212: 563–64.

Caldwell, R. L., 1991. Variation in reproductive behavior in stomatopod crustacea. In *Crustacean Sexual Biology*, ed. R. T. Bauer and J. W. Martin. Columbia University Press.

Campagna, C., B. J. Le Boeuf, and H. L. Cappozzo, 1988. Group raids: A mating strategy of male southern sea lions. *Behaviour* 105: 224–49.

Campbell, R. D., 1974. Cnidaria. In *Reproduction of Marine Invertebrates*. Vol. 1: *Acoelomate and Pseudocoelomate Metazoans*, ed. A. C. Giese and J. S. Pearse. Academic Press.

Carayon, J., 1959. Insémination par "spermalège" et cordon conducteur de spermatozoïdes chez *Stricticimex brevispinosus* Usinger (Heteroptera, Cimicidae). *Revue de zoologie et de botanique africaines* 60: 81–104.

———, 1974. Insémination traumatique hétérosexuelle et homosexuelle chez *Xylocoris maculipennis* (Hem. Anthocoridae). *Comptes rendus de l'Académie des Sciences D, Paris* 278: 2803–06.

Caro, T. M., 1994. *Cheetahs of the Serengeti Plains*. University of Chicago Press.

Carré, D., C. Rouvière, and C. Sardet, 1991. *In vitro* fertilization in ctenophores: Sperm entry, mitosis, and the establishment of bilateral symmetry in *Beroe ovata*. *Developmental Biology* 147: 381–91.

Carré, D., and C. Sardet, 1984. Fertilization and early development in *Beroe ovata*. *Developmental Biology* 105: 188–95.

Carrick, R, and S. E. Ingham, 1962. Studies on the southern elephant seal, *Mirounga leonina* (L.). V. Population dynamics and utilization. *CSIRO Wildlife Research* 7: 198–206.

Carroll, L., 1871. *Through the Looking-Glass, and What Alice Found There.* Macmillan.

Carroll, S. P., and C. Boyd, 1992. Host race radiation in the soapberry bug: Natural history with the history. *Evolution* 46: 1052–69.

Cei, J. M., 1962. *Batracios de Chile*. Ediciones de la Universidad de Chile.

Cerda-Flores, R. M., S. A. Barton, L. F. Marty-Gonzalez, F. Rivas, and R. Chakraborty, 1999. Estimation of nonpaternity in the Mexican population of Nuevo Leon: A validation study with blood group markers. *American Journal of Physical Anthropology* 109: 281–93.

Chao, L., 1992. Evolution of sex in RNA viruses. *Trends in Ecology and Evolution* 7: 147–51.

Chapela, I. H., S. A. Rehner, T. R. Schultz, and U. G. Mueller, 1994. Evolutionary history of the symbiosis between fungus-growing ants and their fungi. *Science* 266: 1691–94.

Chapman, T., 2001. Seminal fluid–mediated fitness traits in *Drosophila*. *Heredity* 87: 511–21.

Charlesworth, D., and B. Charlesworth, 1987. Inbreeding depression and its evolutionary consequences. *Annual Review of Ecology and Systematics* 18: 237–68.

Charnov, E. L., 1982. *The Theory of Sex Allocation*. Princeton University Press.

Charnov, E. L., J. Maynard Smith, and J. J. Bull, 1976. Why be an hermaphrodite? *Nature* 263: 125–26.

Chen, P. S., 1984. The functional morphology and biochemistry of insect male accessory glands and their secretions. *Annual Review of Entomology* 29: 233–55.

Chippindale, A. K., J. R. Gibson, and W. R. Rice, 2001. Negative genetic correlation for adult fitness between sexes reveals ontogenetic conflict in *Drosophila*. *Proceedings of the National Academy of Sciences, USA* 98: 1671–75.

Clutton-Brock, T. H., S. D. Albon, and F. E. Guinness, 1988. Reproductive success in male and female red deer. In *Reproductive Success: Studies of Individual Variation in Contrasting Breeding Systems*, ed. Clutton-Brock. University of Chicago Press.

Cohen, J., 1971. The comparative physiology of gamete populations. *Advances in Comparative Physiology and Biochemistry* 4: 267–380.

Cohen, J., and K. R. Tyler, 1980. Sperm populations in the female genital tract of the rabbit. *Journal of Reproduction and Fertility* 60: 213–18.

Conner, W. E., R. Boada, F. C. Schroeder, A. González, J. Meinwald, and T. Eisner, 2000. Chemical defense: Bestowal of a nuptial alkaloidal garment by a male moth on its mate. *Proceedings of the National Academy of Sciences, USA* 97: 14406–11.

Conover, M. R., and G. L. Hunt, Jr., 1984. Experimental evidence that female-female pairs in gulls result from a shortage of breeding males. *Condor* 86: 472–76.

Cook, D. F., 1990. Differences in courtship, mating, and postcopulatory behaviour between male morphs of the dung beetle *Onthophagus binodis* Thunberg (Coleoptera: Scarabaeidae). *Animal Behaviour* 40: 428–36.

Cook, J. M., S. G. Compton, E. A. Herre, and S. A. West, 1997. Alternative mating tactics and extreme male dimorphism in fig wasps. *Proceedings of the Royal Society of London, B* 264: 747–54.

Córdoba-Aguilar, A., 1999. Male copulatory sensory stimulation induces female ejection of rival sperm in a damselfly. *Proceedings of the Royal Society of London, B* 266: 779–84.

Cosmides, L. M., and J. Tooby, 1981. Cytoplasmic inheritance and intragenomic conflict. *Journal of Theoretical Biology* 89: 83–129.

Cox, C. R., and B. J. Le Boeuf, 1977. Female incitation of male competition: A mechanism in sexual selection. *American Naturalist* 111: 317–35.

Crespi, B. J., 1992. Behavioural ecology of Australian gall thrips (Insecta, Thysanoptera). *Journal of Natural History* 26: 769–809.

Creus, M., J. Balasch, F. Fábregues, J. Martorell, M. Boada, J. Peñarrubia, P. N. Barri, and J. A. Vanrell, 1998. Parental human leukocyte antigens and implantation failure after in-vitro fertilization. *Human Reproduction* 13: 39–43.

Cronin, E. W., Jr., and P. W. Sherman, 1976. A resource-based mating system: The orange-rumped honeyguide. *Living Bird* 15: 5–32.

Cunningham, E. J. A., and K. M. Cheng, 1999. Biases in sperm use in the mallard: No evidence for selection by females based on sperm genotype. *Proceedings of the Royal Society of London, B* 266: 905–10.

Cunningham, E. J. A., and A. F. Russell, 2000. Egg investment is influenced by male attractiveness in the mallard. *Nature* 404: 74–77.

Currie, C. R., U. G. Mueller, and D. Malloch, 1999. The agricultural pathology of ant fungus gardens. *Proceedings of the National Academy of Sciences, USA* 96: 7998–8002.

Curtsinger, J. W., 1991. Sperm competition and the evolution of multiple mating. *American Naturalist* 138: 93–102.

Dallai, R., and B. A. Afzelius, 1984. Paired spermatozoa in *Thermobia* (Insecta, Thysanura). *Journal of Ultrastructure Research* 86: 67–74.

Daly, M., 1978. The cost of mating. *American Naturalist* 112: 771–74.

Darling, F. F., 1963. *A Herd of Red Deer.* Oxford University Press.

Darling, J. D. S., M. L. Noble, and E. Shaw, 1980. Reproductive strategies in the surfperches. I. Multiple insemination in natural populations of the shiner perch, *Cymatogaster aggregata. Evolution* 34: 271–77.

Darwin, C., 1871 (1981). *The Descent of Man, and Selection in Relation to Sex.* (Facsimile). Princeton University Press.

David, P., T. Bjorksten, K. Fowler, and A. Pomiankowski, 2000. Condition-dependent signalling of genetic variation in stalk-eyed flies. *Nature* 406: 186–88.

Davies, J., 1994. Inactivation of antibiotics and the dissemination of resistance genes. *Science* 264: 375–82.

Davies, N. B., 1983. Polyandry, cloaca-pecking, and sperm competition in dunnocks. *Nature* 302: 334–36.

Davis, L. S., F. M. Hunter, R. G. Harcourt, and S. M. Heath, 1998. Reciprocal homosexual mounting in Adélie penguins *Pygoscelis adeliae. Emu* 98: 136–37.

de la Motte, I., and D. Burkhardt, 1983. Portrait of an Asian stalk-eyed fly. *Naturwissenschaften* 70: 451–61.

de Waal, F., 1989. *Peacemaking among Primates.* Harvard University Press.

Decker, M. D., P. G. Parker, D. J. Minchella, and K. N. Rabenold, 1993. Monogamy in black vultures: Genetic evidence from DNA fingerprinting. *Behavioral Ecology* 4: 29–35.

Devine, M. C., 1975. Copulatory plugs in snakes: Enforced chastity. *Science* 187: 844–45.

Dewsbury, D. A., 1982. Ejaculate cost and male choice. *American Naturalist* 119: 601–10.

Dewsbury, D. A., and D. Q. Estep, 1975. Pregnancy in cactus mice: Effects of prolonged copulation. *Science* 187: 552–53.

Diamond, J., 1986. Biology of birds of paradise and bowerbirds. *Annual Review of Ecology and Systematics* 17: 17–37.

———, 1988. Experimental study of bower decoration by the bowerbird *Amblyornis inornatus,* using colored poker chips. *American Naturalist* 131: 631–53.

Dickinson, J. L., 1995. Trade-offs between postcopulatory riding and mate location in the blue milkweed beetle. *Behavorial Ecology* 6: 280–86.

Dickinson, J. L., and R. L. Rutowski, 1989. The function of the mating plug in the chalcedon checkerspot butterfly. *Animal Behaviour* 38: 154–62.

Diesel, R., 1990. Sperm competition and reproductive success in the decapod *Inachus phalangium* (Majidae): A male ghost spider crab that seals off rivals' sperm. *Journal of Zoology* 220: 213–23.

Dixon, A., D. Ross, S. L. C. O'Malley, and T. Burke, 1994. Paternal investment inversely related to degree of extra-pair paternity in the reed bunting. *Nature* 371: 698–700.

Dixson, A. F., 1993. Sexual selection, sperm competition, and the evolution of sperm length. *Folia Primatologica* 61: 221–27.

———, 1998. *Primate Sexuality: Comparative Studies of the Prosimians, Monkeys, Apes, and Human Beings.* Oxford University Press.

Donner, J., 1966. *Rotifers.* Trans. H. G. S. Wright. Frederick Warne.

Downes, J. A., 1978. Feeding and mating in the insectivorous Ceratopogoninae (Diptera). *Memoirs of the Entomological Society of Canada* 104: 1–62.

Drea, C. M., M. L. Weldele, N. G. Forger, E. M. Coscia, L. G. Frank, P. Licht, and S. E. Glickman, 1998. Androgens and masculinization of genitalia in the spotted hyaena (*Crocuta crocuta*). 2. Effects of prenatal anti-androgens. *Journal of Reproduction and Fertility* 113: 117–27.

Duffy, J. E., 1996. Eusociality in a coral-reef shrimp. *Nature* 381: 512–14.

Dunn, D. W., C. S. Crean, and A. S. Gilburn, 2001. Male mating preference for female survivorship in the seaweed fly *Gluma musgravei* (Diptera: Coelopidae). *Proceedings of the Royal Society of London, B* 268: 1255–58.

Dunn, P. O., A. D. Afton, M. L. Gloutney, and R. T. Alisauskas, 1999. Forced copulation results in few extrapair fertilizations in Ross's and lesser snow geese. *Animal Behaviour* 57: 1071–81.

Dusenbery, D. B., 2000. Selection for high gamete encounter rates explains the success of male and female mating types. *Journal of Theoretical Biology* 202: 1–10.

Dybas, L. K., and H. S. Dybas, 1981. Coadaptation and taxonomic differentiation of sperm and spermathecae in featherwing beetles. *Evolution* 35: 168–74.

East, M. L., H. Hofer, and W. Wickler, 1993. The erect "penis" is a flag of submission in a female-dominated society: Greetings in Serengeti spotted hyenas. *Behavioral Ecology and Sociobiology* 33: 355–70.

Eaton, R. L., 1978. Why some felids copulate so much: A model for the evolution of copulation frequency. *Carnivore* 1: 42–51.

Eberhard, W. G., 1980. Evolutionary consequences of intracellular organelle competition. *Quarterly Review of Biology* 55: 231–49.

———, 1985. *Sexual Selection and Animal Genitalia*. Harvard University Press.

———, 1996. *Female Control: Sexual Selection by Cryptic Female Choice*. Princeton University Press.

Edmunds, M., 1988. Sexual cannibalism in mantids. *Trends in Ecology and Evolution* 3: 77.

Edwards, S. V., and P. W. Hedrick, 1998. Evolution and ecology of MHC molecules: From genomics to sexual selection. *Trends in Ecology and Evolution* 13: 305–11.

Eens, M., and R. Pinxten, 1996. Female European starlings increase their copulation solicitation rate when faced with the risk of polygyny. *Animal Behaviour* 51: 1141–47.

Eggert, A.-K., and S. K. Sakaluk, 1994. Sexual cannibalism and its relation to male mating success in sagebrush crickets, *Cyphoderris strepitans* (Haglidae: Orthoptera). *Animal Behaviour* 47: 1171–77.

———, 1995. Female-coerced monogamy in burying beetles. *Behavioral Ecology and Sociobiology* 37: 147–53.

Elder, J. F., Jr., and B. J. Turner, 1995. Concerted evolution of repetitive DNA sequences in eukaryotes. *Quarterly Review of Biology* 70: 297–320.

Elgar, M. A., 1992. Sexual cannibalism in spiders and other invertebrates. In *Cannibalism: Ecology and Evolution among Diverse Taxa*, ed. Elgar and Crespi. Oxford University Press.

Elgar, M. A., and B. J. Crespi, eds., 1992. *Cannibalism: Ecology and Evolution among Diverse Taxa*. Oxford University Press.

Elgar, M. A., and D. R. Nash, 1988. Sexual cannibalism in the garden spider *Araneus diadematus*. *Animal Behaviour* 36: 1511–17.

Emlen, S. T., and P. H. Wrege, 1986. Forced copulations and intra-specific parasitism: Two costs of social living in the white-fronted bee-eater. *Ethology* 71: 2–29.

———, 1994. Gender, status, and family fortunes in the white-fronted bee-eater. *Nature* 367: 129–32.

Enders, R. K., 1952. Reproduction in the mink (*Mustela vison*). *Proceedings of the American Philosophical Society* 96: 691–755.

Enquist, M., and O. Leimar, 1990. The evolution of fatal fighting. *Animal Behaviour* 39: 1–9.

Erwin, J., and T. Maple, 1976. Ambisexual behavior with male-male anal penetration in male rhesus monkeys. *Archives of Sexual Behavior* 5: 9–14.

Estes, R. D., 1991. *The Behavior Guide to African Mammals*. University of California Press.

Evans, H. E., K. M. O'Neill, and R. P. O'Neill, 1986. Nesting site changes and nocturnal clustering in the sand wasp *Bembecinus quinquespinosus* (Hymenoptera: Sphecidae). *Journal of the Kansas Entomological Society* 59: 280–86.

Everitt, D. A., 1981. An ecological study of an Antarctic freshwater pool with particular reference to Tardigrada and Rotifera. *Hydrobiologia* 83: 225–37.

Farr, J. A., 1980. The effects of sexual experience and female receptivity on courtship-rape decisions in male guppies, *Poecilia reticulata* (Pisces: Poeciliidae). *Animal Behaviour* 28: 1195–201.

Faulkes, C. G., and N. C. Bennett, 2001. Family values: Group dynamics and social control of reproduction in African mole-rats. *Trends in Ecology and Evolution* 16: 184–90.

Fietz, J., 1999. Monogamy as a rule rather than exception in nocturnal lemurs: The case of the fat-tailed dwarf lemur, *Cheirogaleus medius*. *Ethology* 105: 259–72.

Fietz, J., H. Zischler, C. Schwiegk, J. Tomiuk, K. H. Dausmann, and J. U. Ganzhorn, 2000. High rates of extra-pair young in the pair-living fat-tailed dwarf lemur, *Cheirogaleus medius*. *Behavioral Ecology and Sociobiology* 49: 8–17.

Fischer, E. A., 1980. The relationship between mating system and simultaneous hermaphroditism in the coral reef fish, *Hypoplectrus nigricans* (Serranidae). *Animal Behaviour* 28: 620–33.

Fisher, R. A., 1999. *The Genetical Theory of Natural Selection: A Complete Variorum Edition*. Oxford University Press.

Flanders, S. E., 1945. The role of spermatophore in the mass propagation of *Macrocentrus ancylivorus*. *Journal of Economic Entomology* 38: 323–27.

Fleming, T. H., S. Maurice, S. L. Buchmann, and M. D. Tuttle, 1994. Reproductive biology and relative male and female fitness in a trioecious cactus, *Pachycereus pringlei* (Cactaceae). *American Journal of Botany* 81: 858–67.

Folgerø, T., K. Bertheussen, S. Lindal, T. Torbergsen, and P. Øian, 1993. Mitochondrial disease and reduced sperm motility. *Human Reproduction* 8: 1863–68.

Foott, J. O., 1970. Nose scars in female sea otters. *Journal of Mammalogy* 51: 621–22.

Forster, L. M., 1992. The stereotyped behaviour of sexual cannibalism in *Latrodectus hasselti* Thorell (Araneae: Theridiidae), the Australian redback spider. *Australian Journal of Zoology* 40: 1–11.

Forsyth, A., and R. D. Montgomerie, 1987. Alternative reproductive tactics in the territorial damselfly *Calopteryx maculata*: Sneaking by older males. *Behavioral Ecology and Sociobiology* 21: 73–81.

Francis, C. M., E. L. P. Anthony, J. A. Brunton, and T. H. Kunz, 1994. Lactation in male fruit bats. *Nature* 367: 691–92.

Frank, L. G., 1997. Evolution of genital masculinization: Why do female hyaenas have such a large "penis"? *Trends in Ecology and Evolution* 12: 58–62.

Frank, L. G., S. E. Glickman, and P. Licht, 1991. Fatal sibling aggression, precocial development, and androgens in neonatal spotted hyenas. *Science* 252: 702–4.

Frank, L. G., M. L. Weldele, and S. E. Glickman, 1995. Masculinization costs in hyaenas. *Nature* 377: 584–85.

Franzén, A., 1977. Sperm structure with regard to fertilization biology and phylogenetics. *Verhandlungen der Deutschen Zoologischen Gesellschaft* 70: 123–38.

Funk, D. H., and D. W. Tallamy, 2000. Courtship role reversal and deceptive signals in the long-tailed dance fly, *Rhamphomyia longicauda*. *Animal Behaviour* 59: 411–21.

Gangrade, G. A., 1963. A contribution to the biology of *Necroscia sparaxes* Westwood (Phasmidae: Phasmida). *The Entomologist* 96: 83–93.

Gary, N. E., 1963. Observations of mating behaviour in the honeybee. *Journal of Apicultural Research* 2: 3–13.

Geist, V., 1971. *Mountain Sheep: A Study in Behavior and Evolution*. University of Chicago Press.

Gerber, H. S., and E. C. Klostermeyer, 1970. Sex control by bees: A voluntary act of egg fertilization during oviposition. *Science* 167: 82–84.

Gharrett, A. J., W. W. Smoker, R. R. Reisenbichler, and S. G. Taylor, 1999. Outbreeding depression in hybrids between odd- and even-broodyear pink salmon. *Aquaculture* 173: 117–29.

Ghiselin, M. T., 1969. The evolution of hermaphroditism among animals. *Quarterly Review of Biology* 44: 189–208.

———, 1974. *The Economy of Nature and the Evolution of Sex*. University of California Press.

Gil, D., J. Graves, N. Hazon, and A. Wells, 1999. Male attractiveness and differential testosterone investment in zebra finch eggs. *Science* 286: 126–28.

Gilbert, A. N., K. Yamazaki, G. K. Beauchamp, and L. Thomas, 1986. Olfactory discrimination of mouse strains (*Mus musculus*) and major histocompatibility types by humans (*Homo sapiens*). *Journal of Comparative Psychology* 100: 262–65.

Gillham, N. W., 1994. *Organelle Genes and Genomes*. Oxford University Press.

Gladstone, D. E., 1979. Promiscuity in monogamous colonial birds. *American Naturalist* 114: 545–57.

Goicoechea, O., O. Garrido, and B. Jorquera, 1986. Evidence for a trophic paternal-larval relationship in the frog *Rhinoderma darwinii*. *Journal of Herpetology* 20: 168–78.

Goldfoot, D. A., H. Westerborg-van Loon, W. Groeneveld, and A. K. Slob, 1980. Behavioral and physiological evidence of sexual climax in the female stumptailed macaque (*Macaca arctoides*). *Science* 208: 1477–79.

Goldstein, I., 2000. Male sexual circuitry. *Scientific American* 283 (August): 56–61.

Gomendio, M., and E. R. S. Roldan, 1991. Sperm competition influences sperm size in mammals. *Proceedings of the Royal Society of London, B* 243: 181–85.

Gonzalez, A., C. Rossini, M. Eisner, and T. Eisner, 1999. Sexually transmitted chemical defense in a moth (*Utetheisa ornatrix*). *Proceedings of the National Academy of Sciences, USA* 96: 5570–74.

Goodall, J., 1986. *The Chimpanzees of Gombe: Patterns of Behavior.* Harvard University Press.

Graffelman, J., and R. F. Hoekstra, 2000. A statistical analysis of the effect of warfare on the human secondary sex ratio. *Human Biology* 72: 433–45.

Gray, D. A., and W. H. Cade, 1999. Sex, death, and genetic variation: Natural and sexual selection on cricket song. *Proceedings of the Royal Society of London, B* 266: 707–09.

Gray, E. M., 1997. Female red-winged blackbirds accrue material benefits from copulating with extra-pair males. *Animal Behaviour* 53: 625–39.

Gross, M. R., and E. L. Charnov, 1980. Alternative male life histories in bluegill sunfish. *Proceedings of the National Academy of Sciences, USA* 77: 6937–40.

Gubernick, D. J., and J. C. Nordby, 1993. Mechanisms of sexual fidelity in the monogamous California mouse, *Peromyscus californicus. Behavioral Ecology and Sociobiology* 32: 211–19.

Gupta, B. L., 1968. Aspects of motility in the non-flagellate spermatozoa of freshwater ostracods. In *Aspects of Cell Motility*, ed. P. L. Miller. Cambridge University Press.

Gust, D. A., and T. P. Gordon, 1991. Male age and reproductive behaviour in sooty mangabeys, *Cercocebus torquatus atys. Animal Behaviour* 41: 277–83.

Gwynne, D. T., 1981. Sexual difference theory: Mormon crickets show role reversal in mate choice. *Science* 213: 779–80.

Gwynne, D. T., and L. W. Simmons, 1990. Experimental reversal of courtship roles in an insect. *Nature* 346: 172–74.

Haddon, M., 1995. Avoidance of post-coital cannibalism in the brachyurid paddle crab *Ovalipes catharus. Oecologia* 104: 256–58.

Hadfield, M. G., 1963. The biology of nudibranch larvae. *Oikos* 14: 85–95.

Haldane, J. B. S., 1949. Disease and evolution. *La Ricerca Scientifica, Supplémento* 19: 68–76.

Hall, J. C., 1994. The mating of a fly. *Science* 264: 1702–14.

Hamer, D. H., S. Hu, V. L. Magnuson, N. Hu, and A. M. L. Pattatucci, 1993. A linkage between DNA markers on the X chromosome and male sexual orientation. *Science* 261: 321–27.

Hamilton, W. D., 1980. Sex versus non-sex versus parasite. *Oikos* 35: 282–90.

———, 1993. Inbreeding in Egypt and in this book: A childish perspective. In *The Natural History of Inbreeding and Outbreeding: Theoretical and Empirical Perspectives*, ed. N. W. Thornhill. University of Chicago Press.

———, 1996. *Narrow Roads of Gene Land*. Vol. 1: *Evolution of Social Behaviour*. W. H. Freeman.

Handford, P., and M. A. Mares, 1985. The mating systems of ratites and tinamous: An evolutionary perspective. *Biological Journal of the Linnean Society* 25: 77–104.

Hanlon, R. T., and J. B. Messenger, 1996. *Cephalopod Behaviour*. Cambridge University Press.

Hansson, B., S. Bensch, and D. Hasselquist, 1997. Infanticide in great reed warblers: Secondary females destroy eggs of primary females. *Animal Behaviour* 54: 297–304.

Harbison, G. R., and R. L. Miller, 1986. Not all ctenophores are hermaphrodites: Studies on the systematics, distribution, sexuality, and development of two species of *Ocyropsis. Marine Biology* 90: 413–24.

Harshman, L. G., and T. Prout, 1994. Sperm displacement without sperm transfer in *Drosophila melanogaster. Evolution* 48: 758–66.

Hartman, D. S., 1979. Ecology and behavior of the manatee (*Trichechus manatus*) in Florida. *Special Publications of the American Society of Mammalogists* 5: 1–153.

Harvey, P. H., and R. M. May, 1989. Out for the sperm count. *Nature* 337: 508–09.

Hastings, I. M., 1992. Population genetic aspects of deleterious cytoplasmic genomes and their effect on the evolution of sexual reproduction. *Genetical Research* 59: 215–25.

Henderson, I. G., P. J. B. Hart, and T. Burke, 2000. Strict monogamy in a semicolonial passerine: The jackdaw *Corvus monedula. Journal of Avian Biology* 31: 177–82.

Hickey, D. A., and M. R. Rose, 1988. The role of gene transfer in the evolution of eukaryotic sex. In *The Evolution of Sex: An Examination of Current Ideas*, ed. R. E. Michod and B. R. Levin. Sinauer.

Hill, W. C. O., 1953. *Primates: Comparative Anatomy and Taxonomy*. Vol. 1: *Strepsirhini*. University of Edinburgh Press.

Hiruki, L. M., W. G. Gilmartin, B. L. Becker, and I. Stirling, 1993. Wounding in Hawaiian monk seals (*Monachus schauinslandi*). *Canadian Journal of Zoology* 71: 458–68.

Hiruki, L. M., I. Stirling, W. G. Gilmartin, T. C. Johanos, and B. L. Becker, 1993. Significance of wounding to female reproductive success in Hawaiian monk seals (*Monachus schauinslandi*) at Laysan Island. *Canadian Journal of Zoology* 71: 469–74.

Hoagland, K. E., 1978. Protandry and the evolution of environmentally-mediated sex change: A study of the mollusca. *Malacologia* 17: 365–91.

Hodgkin, J., and T. M. Barnes, 1991. More is not better: Brood size and population growth in a self-fertilizing nematode. *Proceedings of the Royal Society of London, B* 246: 19–24.

Hoekstra, R. F., 1987. The evolution of sexes. In *The Evolution of Sex and Its Consequences*, ed. S. C. Stearns. Birkhäuser Verlag.

Höglund, J., and R. V. Alatalo, 1995. *Leks.* Princeton University Press.

Höglund, J., R. V. Alatalo, A. Lundberg, P. T. Rintamäki, and J. Lindell, 1999. Microsatellite markers reveal the potential for kin selection on black grouse leks. *Proceedings of the Royal Society of London, B* 266: 813–16.

Holbrook, M. J. L., and J. P. Grassle, 1984. The effect of low density on the development of simultaneous hermaphroditism in male *Capitella* species I (Polychaeta). *Biological Bulletin* 166: 103–09.

Hoogland, J. L., 1998. Why do female Gunnison's prairie dogs copulate with more than one male? *Animal Behaviour* 55: 351–59.

Hoover, J. P., 1998. *Hawai'i's Sea Creatures: A Guide to Hawai'i's Marine Invertebrates.* Mutual Publishing.

Hosken, D. J., and P. I. Ward, 2001. Experimental evidence for testis size evolution via sperm competition. *Ecology Letters* 4: 10–13.

Houck, L. D., 1980. Courtship behavior in the plethodontid salamander, *Desmognathus wrighti. American Zoologist* 20: 825.

Howard, R. D., 1978. The evolution of mating strategies in bullfrogs, *Rana catesbeiana. Evolution* 32: 850–71.

———, 1980. Mating behaviour and mating success in woodfrogs, *Rana sylvatica. Animal Behaviour* 28: 705–16.

Hrdy, S. B., 1979. Infanticide among animals: A review, classification, and examination of the implications for the reproductive strategies of females. *Ethology and Sociobiology* 1: 13–40.

Hu, S., A. M. L. Pattatucci, C. Patterson, L. Li, D. W. Fulker, S. S. Cherny, L. Kruglyak, and D. H. Hamer, 1995. Linkage between sexual orientation and chromosome Xq28 in males but not in females. *Nature Genetics* 11: 248–56.

Huck, U. W., and R. D. Lisk, 1986. Mating-induced inhibition of receptivity in the female golden hamster. *Behavioral and Neural Biology* 45: 107–19.

Hughes, R. L., 1965. Comparative morphology of spermatozoa from five marsupial families. *Australian Journal of Zoology* 13: 533–43.

Hunt, G. L., Jr., A. L. Newman, M. H. Warner, J. C. Wingfield, and J. Kaiwi, 1984. Comparative behavior of male-female and female-female pairs among western gulls prior to egg-laying. *Condor* 86: 157–62.

Hunt, J., and L. W. Simmons, 1998. Patterns of parental provisioning covary with male morphology in a horned beetle (*Onthophagus taurus*) (Coleoptera: Scarabaeidae). *Behavioral Ecology and Sociobiology* 42: 447–51.

Hurst, L. D., 1996. Why are there only two sexes? *Proceedings of the Royal Society of London, B* 263: 415–22.

Hurst, L. D., and W. D. Hamilton, 1992. Cytoplasmic fusion and the nature of sexes. *Proceedings of the Royal Society of London, B* 247: 189–94.

Hurst, L. D., W. D. Hamilton, and R. J. Ladle, 1992. Covert sex. *Trends in Ecology and Evolution* 7: 144–45.

Hutchinson, G. E., 1959. A speculative consideration of certain possible forms of sexual selection in man. *American Naturalist* 93: 81–91.

Hutson, V., and R. Law, 1993. Four steps to two sexes. *Proceedings of the Royal Society of London, B* 253: 43–51.

Insel, T. R., and L. J. Young, 2001. The neurobiology of attachment. *Nature Reviews Neuroscience* 2: 129–36.

Iwasa, Y., and A. Sasaki, 1987. Evolution of the number of sexes. *Evolution* 41: 49–65.

Jaccarini, V., L. Agius, P. J. Schembri, and M. Rizzo, 1983. Sex determination and larval sexual interaction in *Bonellia viridis* Rolando (Echiura: Bonelliidae). *Journal of Experimental Marine Biology and Ecology* 66: 25–40.

Jackson, R. R., and S. D. Pollard, 1997. Jumping spider mating strategies: Sex among cannibals in and out of webs. In *The Evolution of Mating Systems in Insects and Arachnids*, ed. J. C. Choe and B. J. Crespi. Cambridge University Press.

Jamieson, B. G. M., R. Dallai, and B. A. Afzelius, 1999. *Insects: Their Spermatozoa and Phylogeny*. Science Publishers.

Janzen, D. H., 1979a. How many babies do figs pay for babies? *Biotropica* 11: 48–50.

———, 1979b. How to be a fig. *Annual Review of Ecology and Systematics* 10: 13–51.

Jarne, P., and D. Charlesworth, 1993. The evolution of the selfing rate in functionally hermaphrodite plants and animals. *Annual Review of Ecology and Systematics* 24: 441–66.

Jarne, P., M. Vianey-Liaud, and B. Delay, 1993. Selfing and outcrossing in hermaphrodite freshwater gastropods (Basommatophora): Where, when, and why? *Biological Journal of the Linnean Society* 49: 99–125.

Jiggins, F. M., G. D. D. Hurst, and M. E. N. Majerus, 1999. Sex ratio distorting *Wolbachia* causes sex role reversal in its butterfly host. *Proceedings of the Royal Society of London, B* 266: 1–5.

Jivoff, P., 1997. Sexual competition among male blue crab, *Callinectes sapidus*. *Biological Bulletin* 193: 368–80.

Johnson, V. R., Jr., 1969. Behavior associated with pair formation in the banded shrimp *Stenopus hispidus* (Olivier). *Pacific Science* 23: 40–50.

Johnston, M. O., B. Das, and W. R. Hoeh, 1998. Negative correlation between male allocation and rate of self-fertilization in a hermaphroditic animal. *Proceedings of the National Academy of Sciences, USA* 95: 617–20.

Johnstone, R. A., and L. Keller, 2000. How males can gain by harming their mates: Sexual conflict, seminal toxins, and the cost of mating. *American Naturalist* 156: 368–77.

Jones, A. G., S. Östlund-Nilsson, and J. C. Avise, 1998. A microsatellite assessment of sneaked fertilizations and egg thievery in the fifteenspine stickleback. *Evolution* 52: 848–58.

Jones, J. S., and K. E. Wynne-Edwards, 2000. Paternal hamsters mechanically assist the delivery, consume amniotic fluid and placenta, remove fetal membranes, and provide parental care during the birth process. *Hormones and Behavior* 37:116–25.

Judson, O. P., and B. B. Normark, 1996. Ancient asexual scandals. *Trends in Ecology and Evolution* 11: 41–46.

Justine, J.-L., N. le Brun, and X. Mattei, 1985. The aflagellate spermatozoon of *Diplozoon* (Platyhelminthes: Monogenea: Polyopisthocotylea): A demonstrative case of relationship between sperm ultrastructure and biology of reproduction. *Journal of Ultrastructure Research* 92: 47–54.

Kaitala, A., and C. Wiklund, 1994. Polyandrous female butterflies forage for matings. *Behavioral Ecology and Sociobiology* 35: 385–88.

Karr, T. L., and S. Pitnick, 1996. The ins and outs of fertilization. *Nature* 379: 405–06.

Kawano, S., T. Kuroiwa, and R. W. Anderson, 1987. A third multiallelic mating-type locus in *Physarum polycephalum*. *Journal of General Microbiology* 133: 2539–46.

Kawano, S., H. Takano, K. Mori, and T. Kuroiwa, 1991. A mitochondrial plasmid that promotes mitochondrial fusion in *Physarum polycephalum*. *Protoplasma* 160: 167–69.

Keller, L., and H. K. Reeve, 1995. Why do females mate with multiple males? The sexually selected sperm hypothesis. *Advances in the Study of Behavior* 24: 291–315.

Kemp, A., 1995. *The Hornbills*. Oxford University Press.

Kempenaers, B., 1995. Polygyny in the blue tit: Intra- and inter-sexual conflicts. *Animal Behaviour* 49: 1047–64.

Kessel, E. L., 1955. The mating activities of balloon flies. *Systematic Zoology* 4: 97–104.

Kethley, J., 1971. Population regulation in quill mites (Acarina: Syringophilidae). *Ecology* 52: 1113–18.

Kimura, M., 1983. *The Neutral Theory of Molecular Evolution*. Cambridge University Press.

Kjellberg, F., E. Jousselin, J. L. Bronstein, A. Patel, J. Yokoyama, and J.-Y. Rasplus, 2001. Pollination mode in fig wasps: The predictive power of correlated traits. *Proceedings of the Royal Society of London, B* 268: 1113–21.

Kluge, A. G., 1981. The life history, social organization, and parental behavior of *Hyla rosenbergi* Boulenger, a nest-building gladiator frog. *Miscellaneous Publications of the Museum of Zoology, University of Michigan* 160: 1–170.

Koene, J. M., and R. Chase, 1998. Changes in the reproductive system of the snail *Helix aspersa* caused by mucus from the love dart. *Journal of Experimental Biology* 201: 2313–19.

Komers, P. E., and P. N. M. Brotherton, 1997. Female space use is the best predictor of monogamy in mammals. *Proceedings of the Royal Society of London, B* 264: 1261–70.

Kondrashov, A. S., 1984. Deleterious mutations as an evolutionary factor. 1. The advantage of recombination. *Genetical Research* 44: 199–217.

———, 1988. Deleterious mutations and the evolution of sexual reproduction. *Nature* 336: 435–40.

———, 1993. Classification of hypotheses on the advantage of amphimixis. *Journal of Heredity* 84: 372–87.

Koprowski, J. L., 1992. Removal of copulatory plugs by female tree squirrels. *Journal of Mammalogy* 73: 572–76.

Kothe, E., 1996. Tetrapolar fungal mating types: Sexes by the thousands. *FEMS Microbiology Reviews* 18: 65–87.

Kovacs, K. M., and J. P. Ryder, 1983. Reproductive performance of female-female pairs and polygynous trios of ring-billed gulls. *Auk* 100: 658–69.

Krekorian, C., 1976. Field observations in Guyana on the reproductive biology of the spraying characid, *Copeina arnoldi* Regan. *American Midland Naturalist* 96: 88–97.

Kruuk, H., 1972. *The Spotted Hyena: A Study of Predation and Social Behavior.* University of Chicago Press.

Kutschera, U., and P. Wirtz, 1986. Reproductive behaviour and parental care of *Helobdella striata* (Hirudinea, Glossiphoniidae): A leech that feeds its young. *Ethology* 72: 132–42.

Lack, D., 1968. *Ecological Adaptations for Breeding in Birds.* Methuen and Company.

Ladle, R. J., and E. Foster, 1992. Are giant sperm copulatory plugs? *Acta Œcologica* 13: 635–38.

Ladle, R. J., R. A. Johnstone, and O. P. Judson, 1993. Coevolutionary dynamics of sex in a metapopulation: Escaping the Red Queen. *Proceedings of the Royal Society of London, B* 253: 155–60.

Laidlaw, H. H., Jr., and R. E. Page, Jr., 1984. Polyandry in honey bees (*Apis mellifera* L.): Sperm utilization and intracolony genetic relationships. *Genetics* 108: 985–97.

LaMunyon, C. W., and S. Ward, 1998. Larger sperm outcompete smaller sperm in the nematode *Caenorhabditis elegans*. *Proceedings of the Royal Society of London, B* 265: 1997–2002.

Lang, A., 1996. Silk investment in gifts by males of the nuptial feeding spider *Pisaura mirabilis* (Araneae: Pisauridae). *Behaviour* 133: 697–716.

Langlois, T. H., 1965. The conjugal behavior of the introduced European giant garden slug, *Limax maximus* L., as observed on South Bass Island, Lake Erie. *Ohio Journal of Science* 65: 298–304.

Lanier, D. L., D. Q. Estep, and D. A. Dewsbury, 1975. Copulatory behavior of golden hamsters: Effects on pregnancy. *Physiology and Behavior* 15: 209–12.

———, 1979. Role of prolonged copulatory behavior in facilitating reproductive success in a competitive mating situation in laboratory rats. *Journal of Comparative and Physiological Psychology* 93: 781–92.

Lawrence, S. E., 1992. Sexual cannibalism in the praying mantis, *Mantis religiosa*: A field study. *Animal Behaviour* 43: 569–83.

Le Boeuf, B. J., and S. Mesnick, 1990. Sexual behavior of male northern elephant seals: I. Lethal injuries to adult females. *Behaviour* 116: 143–62.

Legendre, R., and A. Lopez, 1974. Étude histologique de quelques formations glandulaires chez les araignées du genre *Argyrodes* (Theridiidae) et description d'un nouveau type de glande: La glande clypéale des males. *Bulletin de la Société Zoologique de France* 99: 453–60.

Lens, L., S. van Dongen, M. van den Broeck, C. van Broeckhoven, and A. A. Dhondt, 1997. Why female crested tits copulate repeatedly with the same partner: Evidence for the mate assessment hypothesis. *Behavioral Ecology* 8: 87–91.

Leonard, J. L., and K. Lukowiak, 1985. Courtship, copulation, and sperm trading in the sea slug, *Navanax inermis* (Opisthobranchia: Cephalaspidae). *Canadian Journal of Zoology* 63: 2719–29.

LeVay, S., 1996. *Queer Science: The Use and Abuse of Research into Homosexuality*. MIT Press.

Levin, B. R., 1988. The evolution of sex in bacteria. In *The Evolution of Sex: An Examination of Current Ideas*, ed. R. E. Michod and B. R. Levin. Sinauer.

Levitan, D. R., and C. Petersen, 1995. Sperm limitation in the sea. *Trends in Ecology and Evolution* 10: 228–31.

Little, T. J., and P. D. N. Hebert, 1996. Ancient asexuals: Scandal or artifact? *Trends in Ecology and Evolution* 11: 296.

Loher, W., I. Ganjian, I. Kubo, D. Stanley-Samuelson, and S. S. Tobe, 1981. Prostaglandins: Their role in egg-laying of the cricket *Teleogryllus commodus*. *Proceedings of the National Academy of Sciences, USA* 78: 7835–38.

Lopez, A., and M. Emerit, 1979. Données complémentaires sur la glande clypéale des *Argyrodes* (Araneae, Theridiidae): Utilisation du microscope électronique à balayage. *Revue Arachnologique* 2: 143–53.

Lott, D. F., 1981. Sexual behavior and intersexual strategies in American bison. *Zeitschrift für Tierpsychologie* 56: 97–114.

Loughry, W. J., P. A. Prodöhl, C. M. McDonough, and J. C. Avise, 1998. Polyembryony in armadillos. *American Scientist* 86: 274–79.

Lovell-Mansbridge, C., and T. R. Birkhead, 1998. Do female pigeons trade pair copulations for protection? *Animal Behaviour* 56: 235–41.

Lowndes, A. G., 1935. The sperms of freshwater Ostracods. *Proceedings of the Zoological Society of London:* 35–48.

Lund, R., 1990. Chondrichthyan life history styles as revealed by the 320 million years old Mississippian of Montana. *Environmental Biology of Fishes* 27: 1–19.

Lutz, R. A., and J. R. Voight, 1994. Close encounter in the deep. *Nature* 371: 563.

Mable, B. K., and S. P. Otto, 1998. The evolution of life cycles with haploid and diploid phases. *BioEssays* 20: 453–62.

Macintyre, S., and A. Sooman, 1991. Non-paternity and prenatal genetic screening. *The Lancet* 338: 869–71.

Mackie, J. B., and M. H. Walker, 1974. A study of the conjugate sperm of the dytiscid water beetles *Dytiscus marginalis* and *Colymbetes fuscus*. *Cell and Tissue Research* 148: 505–19.

MacLeod, J., and R. Z. Gold, 1951. The male factor in fertility and infertility. II. Spermatozoön counts in 1000 men of known fertility and in 1000 cases of infertile marriage. *Journal of Urology* 66: 436–49.

Malo, D., 1903. *Hawaiian Antiquities*. Trans. N. B. Emerson. Hawaiian Gazette.

Mandelbaum, S. L., M. P. Diamond, and A. H. DeCherney, 1987. The impact of antisperm antibodies on human infertility. *Journal of Urology* 138: 1–8.

Mane, S. D., L. Tompkins, and R. C. Richmond, 1983. Male esterase 6 catalyzes the synthesis of a sex pheromone in *Drosophila melanogaster* females. *Science* 222: 419–21.

Mann, T., and C. Lutwak-Mann, 1981. *Male Reproductive Function and Semen: Themes and Trends in Physiology, Biochemistry, and Investigative Andrology*. Springer-Verlag.

Mann, T., A. W. Martin, and J. B. Thiersch, 1966. Spermatophores and spermatophoric reaction in the giant octopus of the North Pacific, *Octopus dofleini martini*. *Nature* 211: 1279–82.

Marcus, E., 1959. Eine neue Gattung der Philinoglossacea. *Kieler Meeresforschungen* 15: 117–19.

Margulis, L., and D. Sagan, 1986. *Origins of Sex: Three Billion Years of Genetic Recombination*. Yale University Press.

Margulis, L., and K. V. Schwartz, 1998. *Five Kingdoms: An Illustrated Guide to the Phyla of Life on Earth*. 3rd edition. W. H. Freeman.

Mark Welch, D., and M. Meselson, 2000. Evidence for the evolution of bdelloid rotifers without sexual reproduction or genetic exchange. *Science* 288: 1211–15.

Markow, T. A., 1982. Mating systems of cactophilic *Drosophila*. In *Ecological Genetics and Evolution: the Cactus-Yeast-Drosophila Model System*, ed. J. S. F. Barker and W. T. Starmer. Academic Press.

———, 1985. A comparative investigation of the mating system of *Drosophila hydei*. *Animal Behaviour* 33: 775–81.

Markow, T. A., M. Quaid, and S. Kerr, 1978. Male mating experience and competitive courtship success in *Drosophila melanogaster*. *Nature* 276: 821–22.

Marks, J. S., J. L. Dickinson, and J. Haydock, 1999. Genetic monogamy in long-eared owls. *Condor* 101: 854–59.

Martan, J., and B. A. Shepherd, 1976. The role of the copulatory plug in reproduction of the guinea pig. *Journal of Experimental Zoology* 196: 79–84.

Mason, L. G., 1980. Sexual selection and the evolution of pair-bonding in soldier beetles. *Evolution* 34: 174–80.

Masters, W. H., and V. E. Johnson, 1966. *Human Sexual Response*. Little, Brown.

Matthews, L. H., 1941. Notes on the genitalia and reproduction of some African bats. *Proceedings of the Zoological Society of London, B* 111: 289–346.

Maynard Smith, J., 1978. *The Evolution of Sex*. Cambridge University Press.

———, 1986. Contemplating life without sex. *Nature* 324: 300–301.

Maynard Smith, J., H. N. Smith, M. O'Rourke, and B. G. Spratt, 1993. How clonal are bacteria? *Proceedings of the National Academy of Sciences, USA* 90: 4384–88.

McBride, A. F., and D. O. Hebb, 1948. Behavior of the captive bottle-nose dolphin, *Tursiops truncatus*. *Journal of Comparative and Physiological Psychology* 41: 111–23.

McComb, K., 1987. Roaring by red deer stags advances the date of oestrus in hinds. *Nature* 330: 648–49.

McCracken, G. F., and P. F. Brussard, 1980. Self-fertilization in the white-lipped land snail *Triodopsis albolabris*. *Biological Journal of the Linnean Society* 14: 429–34.

McCracken, K. G., 2000. The 20-cm spiny penis of the Argentine lake duck (*Oxyura vittata*). *Auk* 117: 820–25.

McDaniel, I. N., and W. R. Horsfall, 1957. Induced copulation of aedine mosquitoes. *Science* 125: 745.

McKaye, K. R., 1983. Ecology and breeding behavior of a cichlid fish, *Cyrtocara eucinostomus*, on a large lek in Lake Malawi, Africa. *Environmental Biology of Fishes* 8: 81–96.

McKaye, K. R., S. M. Louda, and J. R. Stauffer, Jr., 1990. Bower size and male reproductive success in a cichlid fish lek. *American Naturalist* 135: 597–613.

McKinney, F., S. R. Derrickson, and P. Mineau, 1983. Forced copulation in waterfowl. *Behaviour* 86: 250–94.

Mead, A. R., 1942. *The taxonomy, biology, and genital physiology of the giant West Coast land slugs of the genus* Ariolimax *Morch (Gastropoda: Pulmonata)*. Ph.D. thesis, Cornell University.

Meland, S., S. Johansen, T. Johansen, K. Haugli, and F. Haugli, 1991. Rapid disappearance of one parental mitochondrial genotype after isogamous mating in the myxomycete *Physarum polycephalum*. *Current Genetics* 19: 55–60.

Mesnick, S. L., and B. J. Le Boeuf, 1991. Sexual behavior of male northern elephant seals: II. Female response to potentially injurious encounters. *Behaviour* 117: 262–80.

Metz, E. C., and S. R. Palumbi, 1996. Positive selection and sequence rearrangements generate extensive polymorphism in the gamete recognition protein bindin. *Molecular Biology and Evolution* 13: 397–406.

Metz, E. C., R. Robles-Sikisaka, and V. D. Vacquier, 1998. Nonsynonymous substitution in abalone sperm fertilization genes exceeds substitution in introns and mitochondrial DNA. *Proceedings of the National Academy of Sciences, USA* 95: 10676–81.

Michiels, N. K., 1998. Mating conflicts and sperm competition in simultaneous hermaphrodites. In *Sperm Competition and Sexual Selection*, ed. T. R. Birkhead and A. P. Møller. Academic Press.

Michiels, N. K., and L. J. Newman, 1998. Sex and violence in hermaphrodites. *Nature* 391: 647.

Milne, L. J., and M. Milne, 1976. The social behavior of burying beetles. *Scientific American* 235 (August): 84–89.

Mineau, P., and F. Cooke, 1979. Rape in the lesser snow goose. *Behaviour* 70: 280–91.

Mitani, J. C., 1985. Mating behaviour of male orangutans in the Kutai Game Reserve, Indonesia. *Animal Behaviour* 33: 392–402.

Mitas, M., 1997. Trinucleotide repeats associated with human disease. *Nucleic Acids Research* 25: 2245–53.

Møller, A. P., 1989. Ejaculate quality, testes size, and sperm production in mammals. *Functional Ecology* 3: 91–96.

Moore, H. D. M., 1992. Reproduction in the gray short-tailed opossum, *Monodelphis domestica*. In *Reproductive Biology of South American Vertebrates*, ed. W. C. Hamlett. Springer-Verlag.

Moreno, J., L. Boto, J. A. Fargallo, A. de León, and J. Potti, 2000. Absence of extra-pair fertilisations in the chinstrap penguin *Pygoscelis antarctica*. *Journal of Avian Biology* 31: 580–83.

Mori, S., 1995. Factors associated with and fitness effects of nest-raiding in the three-spined stickleback, *Gasterosteus aculeatus*, in a natural situation. *Behaviour* 132: 1011–23.

Morin, J. G., 1986. "Firefleas" of the sea: Luminescent signaling in marine ostracode crustaceans. *Florida Entomologist* 69: 105–21.

Morin, J. G., and A. C. Cohen, 1991. Bioluminescent displays, courtship, and reproduction in ostracodes. In *Crustacean Sexual Biology*, ed. R. T. Bauer and J. W. Martin. Columbia University Press.

Morris, G. K., D. T. Gwynne, D. E. Klimas, and S. K. Sakaluk, 1989. Virgin male mating advantage in a primitive acoustic insect (Orthoptera: Haglidae). *Journal of Insect Behavior* 2: 173–85.

Morton, D. B., and T. D. Glover, 1974. Sperm transport in the female rabbit: The effect of inseminate volume and sperm density. *Journal of Reproduction and Fertility* 38: 139–46.

Moses, M. J., 1961. Spermiogenesis in the crayfish (*Procambarus clarkii*). *Journal of Biophysical and Biochemical Cytology* 9: 222–28.

Moss, C. J., 1983. Oestrous behaviour and female choice in the African elephant. *Behaviour* 86: 167–96.

Müller, H., 1853. Über das Männchen von *Argonauta argo* und die Hectocotylen. *Zeitschrift für Wissenschaftliche Zoologie* 4: 1–35.

Muller, H. J., 1964. The relation of recombination to mutational advance. *Mutation Research* 1: 2–9.

Myers, G. S., 1952. Annual fishes. *Aquarium Journal* 23: 125–41.

Nagelkerke, C. J., and M. W. Sabelis, 1998. Precise control of sex allocation in pseudo-arrhenotokous phytoseiid mites. *Journal of Evolutionary Biology* 11: 649–84.

Nakatsuru, K., and D. L. Kramer, 1982. Is sperm cheap? Limited male fertility and female choice in the lemon tetra (Pisces, Characidae). *Science* 216: 753–55.

Nalepa, C. A., 1988. Reproduction in the woodroach *Cryptocercus punctulatus* Scudder (Dictyoptera: Cryptocercidae): Mating, oviposition, and hatch. *Annals of the Entomological Society of America* 81: 637–41.

Newcomer, S. D., J. A. Zeh, and D. W. Zeh, 1999. Genetic benefits enhance the reproductive success of polyandrous females. *Proceedings of the National Academy of Sciences, USA* 96: 10236–41.

Nishida, T., and K. Kawanaka, 1985. Within-group cannibalism by adult male chimpanzees. *Primates* 26: 274–84.

Nordell, S. E., 1994. Observations of the mating behavior and dentition of the round stingray, *Urolophus halleri*. *Environmental Biology of Fishes* 39: 219–29.

Normark, B. B., 1999. Evolution in a putatively ancient asexual aphid lineage: Recombination and rapid karyotype change. *Evolution* 53: 1458–69.

Nowak, R. M., 1999. *Walker's Mammals of the World*. 6th edition. Johns Hopkins University Press.

Nutting, W. B., 1976. Hair follicle mites (Acari: Demodicidae) of man. *International Journal of Dermatology* 15: 79–98.

O'Neill, K. M., and H. E. Evans, 1981. Predation on conspecific males by females of the beewolf *Philanthus basilaris* Cresson (Hymenoptera: Sphecidae). *Journal of the Kansas Entomological Society* 54: 553–56.

Ober, C., T. Hyslop, S. Elias, L. R. Weitkamp, and W. W. Hauck, 1998. Human leukocyte antigen matching and fetal loss: Results of a 10 year prospective study. *Human Reproduction* 13: 33–38.

Okuda, N., and Y. Yanagisawa, 1996. Filial cannibalism in a paternal mouth-brooding fish in relation to mate availability. *Animal Behaviour* 52: 307–14.

Olsson, M., and T. Madsen, 2001. Promiscuity in sand lizards (*Lacerta agilis*) and adder snakes (*Vipera berus*): Causes and consequences. *Journal of Heredity* 92: 190–97.

Olsson, M., T. Madsen, and R. Shine, 1997. Is sperm really so cheap? Costs of reproduction in male adders, *Vipera berus*. *Proceedings of the Royal Society of London, B* 264: 455–59.

Otronen, M., P. Reguera, and P. I. Ward, 1997. Sperm storage in the yellow dung fly *Scathophaga stercoraria:* Identifying the sperm of competing males in separate female spermathecae. *Ethology* 103: 844–54.

Packer, C., D. A. Gilbert, A. E. Pusey, and S. J. O'Brien, 1991. A molecular genetic analysis of kinship and cooperation in African lions. *Nature* 351: 562–65.

Packer, C., and A. E. Pusey, 1983. Adaptations of female lions to infanticide by incoming males. *American Naturalist* 121: 716–28.

———, 1987. Intrasexual cooperation and the sex ratio in African lions. *American Naturalist* 130: 636–42.

Page, R. E., Jr., 1980. The evolution of multiple mating behavior by honey bee queens (*Apis mellifera* L.). *Genetics* 96: 263–73.

———, 1986. Sperm utilization in social insects. *Annual Review of Entomology* 31: 297–320.

Palumbi, S. R., 1999. All males are not created equal: Fertility differences depend on gamete recognition polymorphisms in sea urchins. *Proceedings of the National Academy of Sciences, USA* 96: 12632–37.

Palumbi, S. R., and E. C. Metz, 1991. Strong reproductive isolation between closely related tropical sea urchins (genus *Echinometra*). *Molecular Biology and Evolution* 8: 227–39.

Pandya, I. J., and J. Cohen, 1985. The leukocytic reaction of the human uterine cervix to spermatozoa. *Fertility and Sterility* 43: 417–21.

Parker, G. A., 1970. Sperm competition and its evolutionary consequences in the insects. *Biological Reviews* 45: 525–67.

———, 1979. Sexual selection and sexual conflict. In *Sexual Selection and Reproductive Competition in Insects*, ed. M. S. Blum and N. A. Blum. Academic Press.

———, 1990a. Sperm competition games: Raffles and roles. *Proceedings of the Royal Society of London, B* 242: 120–26.

———, 1990b. Sperm competition games: Sneaks and extra-pair copulations. *Proceedings of the Royal Society of London, B* 242: 127–33.

Parker, G. A., R. R. Baker, and V. G. F. Smith, 1972. The origin and evolution of gamete dimorphism and the male-female phenomenon. *Journal of Theoretical Biology* 36: 529–53.

Parker, G. A., and L. W. Simmons, 1994. Evolution of phenotypic optima and copula duration in dungflies. *Nature* 370: 53–56.

Partridge, L., 1980. Mate choice increases a component of offspring fitness in fruit flies. *Nature* 283: 290–91.

Pearson, O. P., M. R. Koford, and A. K. Pearson, 1952. Reproduction of the lump-nosed bat (*Corynorhinus rafinesquei*) in California. *Journal of Mammalogy* 33: 273–320.

Pechenik, J. A., and S. Lewis, 2000. Avoidance of drilled gastropod shells by the hermit crab *Pagurus longicarpus* at Nahant, Massachusetts. *Journal of Experimental Marine Biology and Ecology* 253: 17–32.

Peckham, D. J., and A. W. Hook, 1980. Behavioral observations on *Oxybelus* in southeastern North America. *Annals of the Entomological Society of America* 73: 557–67.

Pennings, S. C., 1991. Reproductive behavior of *Aplysia californica* Cooper: Diel patterns, sexual roles and mating aggregations. *Journal of Experimental Marine Biology and Ecology* 149: 249–66.

Petersen, C. W., 1991. Variation in fertilization rate in the tropical reef fish, *Halichoeres bivattatus*: Correlates and implications. *Biological Bulletin* 181: 232–37.

Petrie, M., 1983. Female moorhens compete for small fat males. *Science* 220: 413–15.

———, 1994. Improved growth and survival of offspring of peacocks with more elaborate trains. *Nature* 371: 598–99.

Petrie, M., A. Krupa, and T. Burke, 1999. Peacocks lek with relatives even in the absence of social and environmental cues. *Nature* 401: 155–57.

Phelan, M. C., J. M. Pellock, and W. E. Nance, 1982. Discordant expression of fetal hydantoin syndrome in heteropaternal dizygotic twins. *New England Journal of Medicine* 307: 99–101.

Phillips, D. M., and S. Mahler, 1977. Leukocyte emigration and migration in the vagina following mating in the rabbit. *Anatomical Record* 189: 45–60.

Pilleri, G., M. Gihr, and C. Kraus, 1980. Play behaviour in the Indus and Orinoco dolphin (*Platanista indi* and *Inia geoffrensis*). *Investigations on Cetacea* 11: 57–108.

Pinxten, R., and M. Eens, 1990. Polygyny in the European starling: Effect on female reproductive success. *Animal Behaviour* 40: 1035–47.

Pitnick, S., 1996. Investment in testes and the cost of making long sperm in *Drosophila*. *American Naturalist* 148: 57–80.

Pitnick, S., W. D. Brown, and G. T. Miller, 2001. Evolution of female remating behaviour following experimental removal of sexual selection. *Proceedings of the Royal Society of London, B* 268: 557–63.

Pitnick, S., T. A. Markow, and G. S. Spicer, 1995. Delayed male maturity is a cost of producing large sperm in *Drosophila*. *Proceedings of the National Academy of Sciences, USA* 92: 10614–18.

Pitnick, S., G. S. Spicer, and T. A. Markow, 1995. How long is a giant sperm? *Nature* 375: 109.

Pizzari, T., and T. R. Birkhead, 2000. Female feral fowl eject sperm of subdominant males. *Nature* 405: 787–89.

Plutarch, (n.d.). *The Lives of the Noble Grecians and Romans.* Trans. J. Dryden. Random House.

Polak, M., W. T. Starmer, and J. S. F. Barker, 1998. A mating plug and male mate choice in *Drosophila hibisci* Bock. *Animal Behaviour* 56: 919–26.

Polis, G. A., and R. D. Farley, 1979. Behavior and ecology of mating in the cannibalistic scorpion, *Paruroctonus mesaensis* Stahnke (Scorpionida: Vaejovidae). *Journal of Arachnology* 7: 33–46.

Poole, J. H., 1987. Rutting behavior in African elephants: The phenomenon of musth. *Behaviour* 102: 283–316.

————, 1989a. Announcing intent: The aggressive state of musth in African elephants. *Animal Behaviour* 37: 140–52.

————, 1989b. Mate guarding, reproductive success and female choice in African elephants. *Animal Behaviour* 37: 842–49.

————, 1994. Sex differences in the behaviour of African elephants. In *The Differences between the Sexes,* ed. R. V. Short and E. Balaban. Cambridge University Press.

Poole, J. H., and C. J. Moss, 1981. Musth in the African elephant, *Loxodonta africana. Nature* 292: 830–31.

Potter, D. A., D. L. Wrensch, and D. E. Johnston, 1976. Aggression and mating success in male spider mites. *Science* 193: 160–61.

Pratt, H. L., Jr., 1979. Reproduction in the blue shark, *Prionace glauca. Fishery Bulletin* 77: 445–70.

Proctor, H. C., 1998. Indirect sperm transfer in arthropods: Behavioral and evolutionary trends. *Annual Review of Entomology* 43: 153–74.

Promislow, D. E. L., 1987. Courtship behavior of a plethodontid salamander, *Desmognathus aeneus. Journal of Herpetology* 21: 298–306.

Prowse, N., and L. Partridge, 1997. The effects of reproduction on longevity and fertility in male *Drosophila melanogaster. Journal of Insect Physiology* 43: 501–12.

Pyle, D. W., and M. H. Gromko, 1978. Repeated mating by female *Drosophila melanogaster:* The adaptive importance. *Experientia* 34: 449–50.

————, 1981. Genetic basis for repeated mating in *Drosophila melanogaster. American Naturalist* 117: 133–46.

Radwan, J., 1996. Intraspecific variation in sperm competition success in the bulb mite: A role for sperm size. *Proceedings of the Royal Society of London, B* 263: 855–59.

Randerson, J. P., and L. D. Hurst, 2001. The uncertain evolution of the sexes. *Trends in Ecology and Evolution* 16: 571–79.

Raper, J. R., 1966. *Genetics of Sexuality in Higher Fungi.* Ronald Press.

Réale, D., P. Boussès, and J.-L. Chapuis, 1996. Female-biased mortality induced by male sexual harassment in a feral sheep population. *Canadian Journal of Zoology* 74: 1812–18.

Rees, E. C., P. Lievesley, R. A. Pettifor, and C. Perrins, 1996. Mate fidelity in swans: An interspecific comparison. In *Partnerships in Birds: The Study of Monogamy*, ed. J. M. Black. Oxford University Press.

Reish, D. J., 1957. The life history of the polychaetous annelid *Neanthes caudata* (delle Chiaje), including a summary of development in the family Nereidae. *Pacific Science* 11: 216–28.

Reiswig, H. M., 1970. Porifera: Sudden sperm release by tropical demospongiae. *Science* 170: 538–39.

Renner, S. S., and R. E. Ricklefs, 1995. Dioecy and its correlates in the flowering plants. *American Journal of Botany* 82: 596–606.

Ribble, D. O., 1991. The monogamous mating system of *Peromyscus californicus* as revealed by DNA fingerprinting. *Behavioral Ecology and Sociobiology* 29: 161–66.

Ricci, C., 1992. Rotifera: Parthenogenesis and heterogony. In *Sex Origin and Evolution*, ed. R. Dallai. Mucchi.

———, 1987. Ecology of bdelloids: How to be successful. *Hydrobiologia* 147: 117–27.

Rice, G., C. Anderson, N. Risch, and G. Ebers, 1999. Male homosexuality: Absence of linkage to microsatellite markers at Xq28. *Science* 284: 665–67.

Rice, W. R., 1984. Sex chromosomes and the evolution of sexual dimorphism. *Evolution* 38: 735–42.

———, 1996. Sexually antagonistic male adaptation triggered by experimental arrest of female evolution. *Nature* 381: 232–34.

Rice, W. R., and E. E. Hostert, 1993. Laboratory experiments on speciation: What have we learned in 40 years? *Evolution* 47: 1637–53.

Ridley, M., 1988. Mating frequency and fecundity in insects. *Biological Reviews* 63: 509–49.

Ridley, M., and C. Rechten, 1981. Female sticklebacks prefer to spawn with males whose nests contain eggs. *Behaviour* 76: 152–61.

Ridley, M. W., and D. A. Hill, 1987. Social organization in the pheasant (*Phasianus colchicus*): Harem formation, mate selection, and the role of mate guarding. *Journal of Zoology* 211: 619–30.

Riedman, M. L., and J. A. Estes, 1990. The sea otter (*Enhydra lutris*): Behavior, ecology, and natural history. *Biological Report of the U.S. Fish and Wildlife Service* 90: 1–126.

Riemann, J. G., D. J. Moen, and B. J. Thorson, 1967. Female monogamy and its control in houseflies. *Journal of Insect Physiology* 13: 407–18.

Rijksen, H. D., 1978. *A Field Study on Sumatran Orang Utans* (Pongo pygmaeus abelii *Lesson 1827*): *Ecology, Behaviour, and Conservation*. H. Veenman and Zonen B. V.

Rissing, S. W., and G. B. Pollock, 1987. Queen aggression, pleometrotic advantage, and brood raiding in the ant *Veromessor pergandei* (Hymenoptera: Formicidae). *Animal Behaviour* 35: 975–81.

Robbins, M. M., 1999. Male mating patterns in wild multimale mountain gorilla groups. *Animal Behaviour* 57: 1013–20.

Robinson, M. H., 1982. Courtship and mating behavior in spiders. *Annual Review of Entomology* 27: 1–20.

Robinson, M. H., and B. Robinson, 1980. Comparative studies of the courtship and mating behavior of tropical araneid spiders. *Pacific Insects Monograph* 36: 1–218.

Roeder, K. D., 1935. An experimental analysis of the sexual behavior of the praying mantis (*Mantis religiosa* L.). *Biological Bulletin* 69: 203–20.

Rogers, D. W., and R. Chase, 2001. Dart receipt promotes sperm storage in the garden snail *Helix aspersa*. *Behavioral Ecology and Sociobiology* 50: 122–27.

Rohwer, S., 1978. Parent cannibalism of offspring and egg raiding as a courtship strategy. *American Naturalist* 112: 429–40.

Roldan, E. R. S., M. Gomendio, and A. D. Vitullo, 1992. The evolution of eutherian spermatozoa and underlying selective forces: Female selection and sperm competition. *Biological Reviews* 67: 551–93.

Romero, G. A., and C. E. Nelson, 1986. Sexual dimorphism in *Catasetum* orchids: Forcible pollen emplacement and male flower competition. *Science* 232: 1538–40.

Ross, P., Jr., and D. Crews, 1977. Influence of the seminal plug on mating behaviour in the garter snake. *Nature* 267: 344–45.

Rothschild, Lord, 1961. Structure and movements of tick spermatozoa (Arachnida, Acari). *Quarterly Journal of Microscopical Science* 102: 239–47.

Rovner, J. S., 1972. Copulation in the lycosid spider (*Lycosa rabida* Walckenaer): A quantitive study. *Animal Behaviour* 20: 133–38.

Rowley, I., and E. Russell, 1990. "Philandering"—A mixed mating strategy in the splendid fairy-wren *Malurus splendens*. *Behavioral Ecology and Sociobiology* 27: 431–37.

Russell, E., and I. Rowley, 1996. Partnerships in promiscuous splendid fairy-wrens. In *Partnerships in Birds: The Study of Monogamy*, ed. J. M. Black. Oxford University Press.

Rutowski, R. L., 1983. Mating and egg mass production in the aeolid nudibranch *Hermissenda crassicornis* (Gastropoda: Opisthobranchia). *Biological Bulletin* 165: 276–85.

Ryner, L. C., S. F. Goodwin, D. H. Castrillon, A. Anand, A. Villella, B. S. Baker, J. C. Hall, B. J. Taylor, and S. A. Wasserman, 1996. Control of male sexual behavior and sexual orientation in *Drosophila* by the *fruitless* gene. *Cell* 87: 1079–89.

Saito, T., and K. Konishi, 1999. Direct development in the sponge-associated deep-sea shrimp *Spongicola japonica* (Decapoda: Spongicolidae). *Journal of Crustacean Biology* 19: 46–52.

Sakaluk, S. K., P. J. Bangert, A.-K. Eggert, C. Gack, and L. V. Swanson, 1995. The gin trap as a device facilitating coercive mating in sagebrush crickets. *Proceedings of the Royal Society of London, B* 261: 65–71.

Sakaluk, S. K., and J. J. Belwood, 1984. Gecko phonotaxis to cricket calling song: A case of satellite predation. *Animal Behaviour* 32: 659–62.

Sandell, M. I., and H. G. Smith, 1997. Female aggression in the European starling during the breeding season. *Animal Behaviour* 53: 13–23.

Sasaki, T., and O. Iwahashi, 1995. Sexual cannibalism in an orb-weaving spider *Argiope aemula. Animal Behaviour* 49: 1119–21.

Sasse, G., H. Müller, R. Chakraborty, and J. Ott, 1994. Estimating the frequency of nonpaternity in Switzerland. *Human Heredity* 44: 337–43.

Sato, T., 1994. Active accumulation of spawning substrate: A determinant of extreme polygyny in a shell-brooding cichlid fish. *Animal Behaviour* 48: 669–78.

Schal, C., and W. J. Bell, 1982. Ecological correlates of paternal investment of urates in a tropical cockroach. *Science* 218: 170–73.

Schmidt, A. M., L. A. Nadal, M. J. Schmidt, and N. B. Beamer, 1979. Serum concentrations of oestradiol and progesterone during the normal oestrous cycle and early pregnancy in the lion (*Panthera leo*). *Journal of Reproduction and Fertility* 57: 267–72.

Seemanová, E., 1971. A study of children of incestuous matings. *Human Heredity* 21: 108–28.

Sever, Z., and H. Mendelssohn, 1988. Copulation as a possible mechanism to maintain monogamy in porcupines, *Hystrix indica. Animal Behaviour* 36: 1541–42.

Shapiro, D. Y., A. Marconato, and T. Yoshikawa, 1994. Sperm economy in a coral reef fish, *Thalassoma bifasciatum. Ecology* 75: 1334–44.

Shapiro, L. E., and D. A. Dewbury, 1990. Differences in affiliative behavior, pair bonding, and vaginal cytology in two species of vole (*Microtus ochrogaster* and *M. montanus*). *Journal of Comparative Psychology* 104: 268–74.

Sharma, R. P., 1977. Light-dependent homosexual activity in males of a mutant of *Drosophila melanogaster. Experientia* 33: 171–73.

Shaw, E., and J. Allen, 1977. Reproductive behavior in the female shiner perch *Cymatogaster aggregata. Marine Biology* 40: 81–86.

Shellman-Reeve, J. S., 1999. Courting strategies and conflicts in a monogamous, biparental termite. *Proceedings of the Royal Society of London, B* 266: 137–44.

Shepher, J., 1971. Mate selection among second-generation kibbutz adolescents and adults: Incest avoidance and negative imprinting. *Archives of Sexual Behavior* 1: 293–307.

Sherman, P. W., 1989. Mate guarding as paternity insurance in Idaho ground squirrels. *Nature* 338: 418–20.

Short, R. V., 1979. Sexual selection and its component parts, somatic and genital selection, as illustrated by man and the great apes. *Advances in the Study of Behavior* 9: 131–58.

Shuster, S. M., 1989. Male alternative reproductive strategies in a marine isopod crustacean (*Paracerceis sculpta*): The use of genetic markers to measure differences in fertilization success among α-, β-, and γ-males. *Evolution* 43: 1683–98.

————, 1990. Courtship and female mate selection in a marine isopod crustacean, *Paracerceis sculpta*. *Animal Behaviour* 40: 390–99.

————, 1991. The ecology of breeding females and the evolution of polygyny in *Paracerceis sculpta*, a marine isopod crustacean. In *Crustacean Sexual Biology*, ed. R. T. Bauer and J. W. Martin. Columbia University Press.

Shuster, S. M., and M. J. Wade, 1991. Equal mating success among male reproductive strategies in a marine isopod. *Nature* 350: 608–10.

Simmons, L. W., 1987. Female choice contributes to offspring fitness in the field cricket, *Gryllus bimaculatus* (De Geer). *Behavioral Ecology and Sociobiology* 21: 313–21.

Simmons, L. W., and M. T. Siva-Jothy, 1998. Sperm competition in insects: Mechanisms and the potential for selection. In *Sperm Competition and Sexual Selection*, ed. T. R. Birkhead and A. P. Møller. Academic Press.

Simmons, L. W., P. Stockley, R. L. Jackson, and G. A. Parker, 1996. Sperm competition or sperm selection: No evidence for female influence over paternity in yellow dung flies *Scatophaga stercoraria*. *Behavioral Ecology and Sociobiology* 38: 199–206.

Simmons, L. W., J. L. Tomkins, and J. Hunt, 1999. Sperm competition games played by dimorphic male beetles. *Proceedings of the Royal Society of London*, B 266: 145–50.

Simmons, R. E., 1988. Food and the deceptive acquisition of mates by polygynous male harriers. *Behavorial Ecology and Sociobiology* 23: 83–92.

Sinervo, B., and C. M. Lively, 1996. The rock-paper-scissors game and the evolution of alternative male strategies. *Nature* 380: 240–43.

Siniff, D. B., I. Stirling, J. L. Bengston, and R. A. Reichle, 1979. Social and reproductive behavior of crabeater seals (*Lobodon carcinophagus*) during the austral spring. *Canadian Journal of Zoology* 57: 2243–55.

Slagsvold, T., T. Amundsen, S. Dale, and H. Lampe, 1992. Female-female aggression explains polyterritoriality in male pied flycatchers. *Animal Behaviour* 43: 397–407.

Slagsvold, T., and J. T. Lifjeld, 1994. Polygyny in birds: The role of competition between females for male parental care. *American Naturalist* 143: 59–94.

Slob, A. K., W. H. Groeneveld, and J. J. van der Werff Ten Bosch, 1986. Physiological changes during copulation in male and female stumptail macaques (*Macaca arctoides*). *Physiology and Behavior* 38: 891–95.

Smith, R. L., 1979. Repeated copulation and sperm precedence: Paternity assurance for a male brooding water bug. *Science* 205: 1029–31.

Smuts, B. B., and R. W. Smuts, 1993. Male aggression and sexual coercion of females in nonhuman primates and other mammals: Evidence and theoretical implications. *Advances in the Study of Behavior* 22: 1–63.

Smuts, B. B., and J. M. Watanabe, 1990. Social relationships and ritualized greetings in adult male baboons (*Papio cynocephalus anubis*). *International Journal of Primatology* 11: 147–72.

Solymar, B. D., and W. H. Cade, 1990. Heritable variation for female mating frequency in field crickets, *Gryllus integer*. *Behavioral Ecology and Sociobiology* 26: 73–76.

Sommer, V., and U. Reichard, 2000. Rethinking monogamy: The gibbon case. In *Primate Males: Causes and Consequences of Variation in Group Composition*, ed. P. M. Kappeler. Cambridge University Press.

Spinka, M., 1988. Different outcomes of sperm competition in right and left sides of the female reproductive tract revealed by thymidine-³H-labeled spermatozoa in the rat. *Gamete Research* 21: 313–21.

Springer, S., 1948. Oviphagous embryos of the sand shark, *Carcharias taurus*. *Copeia*: 153–56.

Staedler, M., and M. Riedman, 1993. Fatal mating injuries in female sea otters (*Enhydra lutris nereis*). *Mammalia* 57: 135–39.

Starks, P. T., and E. S. Poe, 1997. "Male-stuffing" in wasp societies. *Nature* 389: 450.

Steinkraus, D. C., and E. A. Cross, 1993. Description and life history of *Acarophenax mahunkai*, n. sp. (Acari, Tarsonemina: Acarophenacidae), an egg parasite of the lesser mealworm (Coleoptera: Tenebrionidae). *Annals of the Entomological Society of America* 86: 239–49.

Stevens, J. D., 1974. The occurrence and significance of tooth cuts on the blue shark (*Prionace glauca* L.) from British waters. *Journal of the Marine Biological Association of the United Kingdom* 54: 373–78.

Stockley, P., M. J. G. Gage, G. A. Parker, and A. P Møller, 1996. Female reproductive biology and the coevolution of ejaculate characteristics in fish. *Proceedings of the Royal Society of London, B* 263: 451–58.

Stone, G. N., 1995. Female foraging responses to sexual harassment in the solitary bee *Anthophora plumipes*. *Animal Behaviour* 50: 405–12.

Stone, G. N., P. M. Loder, and T. M. Blackburn, 1995. Foraging and courtship behaviour in males of the solitary bee *Anthophora plumipes* (Hymenoptera: Anthophoridae): Thermal physiology and the roles of body size. *Ecological Entomology* 20: 169–83.

Stoner, D. S., and I. L. Weissman, 1996. Somatic and germ cell parasitism in a colonial ascidian: Possible role for a highly polymorphic allorecognition system. *Proceedings of the National Academy of Sciences, USA* 93: 15254–59.

Suetonius, 1957. *The Twelve Caesars*. Trans. R. Graves. Penguin Books.

Summers, K., 1989. Sexual selection and intra-female competition in the green poison-dart frog, *Dendrobates auratus*. *Animal Behaviour* 37: 797–805.

Svensson, B. G., E. Petersson, and M. Frisk, 1990. Nuptial gift size prolongs copulation duration in the dance fly *Empis borealis*. *Ecological Entomology* 15: 225–29.

Swanson, W. J., C. F. Aquadro, and V. D. Vacquier, 2001. Polymorphism in abalone fertilization proteins is consistent with the neutral evolution of the egg's receptor for lysin (VERL) and positive Darwinian selection of sperm lysin. *Molecular Biology and Evolution* 18: 376–83.

Swanson, W. J., and V. D. Vacquier, 1997. The abalone egg vitelline envelope receptor for sperm lysin is a giant multivalent molecule. *Proceedings of the National Academy of Sciences, USA* 94: 6724–29.

———, 1998. Concerted evolution in an egg receptor for a rapidly evolving abalone sperm protein. *Science* 281: 710–12.

———, 2002. The rapid evolution of reproductive proteins. *Nature Reviews Genetics* 3: 137–44.

Swanson, W. J., Z. Yang, M. F. Wolfner, and C. F. Aquadro, 2001. Positive Darwinian selection drives the evolution of several female reproductive proteins in mammals. *Proceedings of the National Academy of Sciences, USA* 98: 2509–14.

Sykes, B., and C. Irven, 2000. Surnames and the Y chromosome. *American Journal of Human Genetics* 66: 1417–19.

Sylvestre, J.-P., 1985. Some observations on behaviour of two Orinoco dolphins (*Inia geoffrensis humboldtiana*, Pilleri and Gihr 1977), in captivity, at Duisburg Zoo. *Aquatic Mammals* 11: 58–65.

Synnott, A. L., W. J. Fulkerson, D. R. Lindsay, 1981. Sperm output by rams and distribution amongst ewes under conditions of continual mating. *Journal of Reproduction and Fertility* 61: 355–61.

Taborsky, M., 1994. Sneakers, satellites, and helpers: Parasitic and cooperative behavior in fish reproduction. *Advances in the Study of Behavior* 23: 1–100.

Tarpy, D. R., and R. E. Page, Jr., 2001. The curious promiscuity of queen honey bees (*Apis mellifera*): Evolutionary and behavioral mechanisms. *Annales Zoologici Fennici* 38: 255–65.

Tavolga, M. C., 1966. Behavior of the bottlenose dolphin (*Tursiops truncatus*): Social interactions in a captive colony. In *Whales, Dolphins, and Porpoises*, ed. K. S. Norris. University of California Press.

Taylor, D. S., 1990. Adaptive specializations of the cyprinodont fish *Rivulus marmoratus*. *Florida Scientist* 53: 239–48.

———, 2001. Physical variability and fluctuating asymmetry in heterozygous and homozygous populations of *Rivulus marmoratus*. *Canadian Journal of Zoology* 79: 766–78.

Taylor, D. S., M. T. Fisher, and B. J. Turner, 2001. Homozygosity and heterozygosity in three populations of *Rivulus marmoratus*. *Environmental Biology of Fishes* 61: 455–59.

Taylor, O. R., Jr., 1967. Relationship of multiple mating to fertility in *Atteva punctella* (Lepidoptera: Yponomeutidae). *Annals of the Entomological Society of America* 60: 583–90.

Taylor, V. A., B. M. Luke, and M. B. Lomas, 1982. The giant sperm of a minute beetle. *Tissue and Cell* 14: 113–23.

Temerlin, M. K., 1976. *Lucy: Growing Up Human.* Souvenir Press.

Thomas, D. W., M. B. Fenton, and R. M. R. Barclay, 1979. Social behavior of the little brown bat, *Myotis lucifugus* I. Mating behavior. *Behavioral Ecology and Sociobiology* 6: 129–36.

Thompson, T. E., 1973. Euthyneuran and other molluscan spermatozoa. *Malacologia* 14: 167–206.

Thornhill, R., 1975. Scorpionflies as kleptoparasites of web-building spiders. *Nature* 258: 709–11.

———, 1976. Sexual selection and nuptial feeding behavior in *Bittacus apicalis* (Insecta: Mecoptera). *American Naturalist* 110: 529–48.

———, 1980. Rape in *Panorpa* scorpionflies and a general rape hypothesis. *Animal Behaviour* 28: 52–59.

———, 1988. The jungle fowl hen's cackle incites male competition. *Verhandlungen der Deutschen Zoologischen Gesellschaft* 81: 145–54.

Tinklepaugh, O. L., 1930. Occurrence of vaginal plug in a chimpanzee. *Anatomical Record* 46: 329–32.

Tolson, P. J., 1992. The reproductive biology of the neotropical boid genus *Epicrates* (Serpentes: Boidae). In *Reproductive Biology of South American Vertebrates*, ed. W. C. Hamlett. Springer-Verlag.

Tompa, A. S., 1984. Land snails (Stylommatophora). In *The Mollusca*. Vol. 7: *Reproduction*, ed. A. S. Tompa, N. H. Verdonk, and J. A. M. van den Biggelaar. Academic Press.

Tooby, J., 1982. Pathogens, polymorphism, and the evolution of sex. *Journal of Theoretical Biology* 97: 557–76.

Treat, A. E., 1965. Sex-distinctive chromatin and the frequency of males in the moth ear mite. *New York Entomological Society* 73: 12–18.

———, 1975. *Mites of Moths and Butterflies.* Cornell University Press.

Trivers, R. L., 1972. Parental investment and sexual selection. In *Sexual Selection and the Descent of Man, 1871–1971*, ed. B. Campbell. Heinemann.

Trumbo, S. T., and A. J. Fiore, 1994. Interspecific competition and the evolution of communal breeding in burying beetles. *American Midland Naturalist* 131: 169–74.

Tuttle, E. M., S. Pruett-Jones, and M. S. Webster, 1996. Cloacal protuberances and extreme sperm production in Australian fairy-wrens. *Proceedings of the Royal Society of London, B* 263: 1359–64.

Tuttle, M. D., and M. J. Ryan, 1981. Bat predation and the evolution of frog vocalizations in the Neotropics. *Science* 214: 677–78.

Tyldesley, J., 1994. *Daughters of Isis: Women of Ancient Egypt.* Viking.

Tyndale-Biscoe, M., 1996. Australia's introduced dung beetles: Original releases and redistributions. *CSIRO Divison of Entomology Technical Report* 62: 1–149.

Vahed, K., 1998. The function of nuptial feeding in insects: A review of empirical studies. *Biological Reviews* 73: 43–78.

van den Berghe, P. L., and G. M. Mesher, 1980. Royal incest and inclusive fitness. *American Ethnologist* 7: 300–17.

van der Lande, V. M., and R. C. Tinsley, 1976. Studies on the anatomy, life history, and behaviour of *Marsupiobdella africana* (Hirudinea: Glossiphoniidae). *Journal of Zoology* 180: 537–63.

Vasey, P. L., 1998. Female choice and inter-sexual competition for female sexual partners in Japanese macaques. *Behaviour* 135: 579–97.

Veiga, J. P., 1990. Sexual conflict in the house sparrow: Interference between polygynously mated females versus asymmetric male investment. *Behavioral Ecology and Sociobiology* 27: 345–50.

Verrell, P. A., 1986. Limited male mating capacity in the smooth newt, *Triturus vulgaris vulgaris* (Amphibia). *Journal of Comparative Psychology* 100: 291–95.

Verrell, P. A., T. R. Halliday, and M. L. Griffiths, 1986. The annual reproductive cycle of the smooth newt (*Triturus vulgaris*) in England. *Journal of Zoology, A* 210: 101–19.

vom Saal, F. S., 1989. Sexual differentiation in litter-bearing mammals: Influence of sex of adjacent fetuses *in utero*. *Journal of Animal Science* 67: 1824–40.

Voss, R., 1979. Male accessory glands and the evolution of copulatory plugs in rodents. *Occasional Papers of the Museum of Zoology, University of Michigan* 689: 1–27.

Vreys, C., and N. K Michiels, 1998. Sperm trading by volume in a hermaphroditic flatworm with mutual penis intromission. *Animal Behaviour* 56: 777–85.

Waage, J. K., 1979. Dual function of the damselfly penis: Sperm removal and transfer. *Science* 203: 916–18.

Waddell, D. R., 1992. Cannibalism in lower eukaryotes. In *Cannibalism: Ecology and Evolution among Diverse Taxa*, ed. M. A. Elgar and B. J. Crespi. Oxford University Press.

Waddy, S. L., and D. E. Aiken, 1991. Mating and insemination in the American lobster, *Homarus americanus*. In *Crustacean Sexual Biology*, ed. R. T. Bauer and J. W. Martin. Columbia University Press.

Wagner, R. H., 1996. Male-male mountings by a sexually monomorphic bird: Mistaken identity or fighting tactic? *Journal of Avian Biology* 27: 209–14.

Waights, V., 1996. Female sexual interference in the smooth newt, *Triturus vulgaris vulgaris*. *Ethology* 102: 736–47.

Walker, W. F., 1980. Sperm utilization strategies in nonsocial insects. *American Naturalist* 115: 780–99.

Wallach, S. J. R., and B. L. Hart, 1983. The role of the striated penile muscles of the male rat in seminal plug dislodgement and deposition. *Physiology and Behavior* 31: 815–21.

Walter, B., and F. Trillmich, 1994. Female aggression and male peace-keeping in a cichlid fish harem: Conflict between and within the sexes in *Lamprologus ocellatus. Behavioral Ecology and Sociobiology* 34: 105–12.

Walter, D. E., and H. C. Proctor, 1999. *Mites: Ecology, Evolution, and Behaviour.* CABI Publishing.

Ward, P. I., 1998. A possible explanation for cryptic female choice in the yellow dung fly, *Scathophaga stercoraria* (L.). *Ethology* 104: 97–110.

Ward, S., and J. S. Carrel, 1979. Fertilization and sperm competition in the nematode *Caenorhabditis elegans. Developmental Biology* 73: 304–21.

Warner, R. R., and E. T. Schultz, 1992. Sexual selection and male characteristics in the bluehead wrasse, *Thalassoma bifasciatum:* Mating site acquisition, mating site defense, and female choice. *Evolution* 46: 1421–42.

Warner, R. R., D. Y. Shapiro, A. Marcanato, and C. W. Petersen, 1995. Sexual conflict: Males with highest mating success convey the lowest fertilization benefits to females. *Proceedings of the Royal Society of London, B* 262: 135–39.

Wassersug, R., 1997. Wrapping the armadillo's penis. *Nature* 388: 826–27.

Watanabe, S., M. Hara, and Y. Watanabe, 2000. Male internal fertilization and introsperm-like sperm of the seaweed pipefish (*Syngnathus schlegeli*). *Zoological Science* 17: 759–67.

Webster, M. S., 1991. Male parental care and polygyny in birds. *American Naturalist* 137: 274–80.

Wedekind, C., and S. Füri, 1997. Body odour preferences in men and women: Do they aim for specific MHC combinations or simply heterozygosity? *Proceedings of the Royal Society of London, B* 264: 1471–79.

Wedekind, C., T. Seebeck, F. Bettens, and A. J. Paepke, 1995. MHC-dependent mate preferences in humans. *Proceedings of the Royal Society of London, B* 260: 245–49.

Wei, Y.-H., and S.-H. Kao, 2000. Mitochondrial DNA mutation and depletion are associated with decline of fertility and motility of human sperm. *Zoological Studies* 39: 1–12.

Weiss, M. J., and D. P. Levy, 1979. Sperm in "parthenogenetic" freshwater gastrotrichs. *Science* 205: 302–03.

Wells, K. D., 1978. Courtship and parental behavior in a Panamanian poison-arrow frog (*Dendrobates auratus*). *Herpetologica* 34: 148–55.

Werren, J. H., 1980. Sex ratio adaptations to local mate competition in a parasitic wasp. *Science* 208: 1157–59.

———, 1993. The evolution of inbreeding in haplodiploid organisms. In *The Natural History of Inbreeding and Outbreeding: Theoretical and Empirical Perspectives*, ed. N. W. Thornhill. University of Chicago Press.

West, S. A., E. A. Herre, D. M. Windsor, and P. R. S. Green, 1996. The ecology and evolution of the New World non-pollinating fig wasp communities. *Journal of Biogeography* 23: 447–58.

West, S. A., M. G. Murray, C. A. Machado, A. S. Griffin, and E. A. Herre, 2001. Testing Hamilton's rule with competition between relatives. *Nature* 409: 510–13.

Wheeler, W. M., 1928. *The Social Insects: Their Origin and Evolution.* Harcourt Brace.

White, M. J. D., 1973. *Animal Cytology and Evolution.* 3rd edition. Cambridge University Press.

Wiebes, J. T., 1979. Co-evolution of figs and their insect pollinators. *Annual Review of Ecology and Systematics* 10: 1–12.

Wiese, L., W. Wiese, and D. A. Edwards, 1979. Inducible anisogamy and the evolution of oogamy from isogamy. *Annals of Botany* 44: 131–39.

Wikelski, M., and S. Bäurle, 1996. Pre-copulatory ejaculation solves time constraints during copulations in marine iguanas. *Proceedings of the Royal Society of London, B* 263: 439–44.

Wilkinson, G. S., and P. R. Reillo, 1994. Female choice response to artificial selection on an exaggerated male trait in a stalk-eyed fly. *Proceedings of the Royal Society of London, B* 255: 1–6.

Williams, G. C., 1966. *Adaptation and Natural Selection: A Critique of Some Current Evolutionary Thought.* Princeton University Press.

———, 1975. *Sex and Evolution.* Princeton University Press.

Wilson, E. O., 1971. *The Insect Societies.* Harvard University Press.

———, 1975. *Sociobiology: The New Synthesis.* Harvard University Press.

Wilson, J. R., N. Adler, and B. Le Boeuf, 1965. The effects of intromission frequency on successful pregnancy in the female rat. *Proceedings of the National Academy of Sciences, USA* 53: 1392–95.

Wilson, N., S. C. Tubman, P. E. Eady, and G. W. Robertson, 1997. Female genotype affects male success in sperm competition. *Proceedings of the Royal Society of London, B* 264: 1491–95.

Winslow, J. T., N. Hastings, C. S. Carter, C. R. Harbaugh, and T. R. Insel, 1993. A role for central vasopressin in pair bonding in monogamous prairie voles. *Nature* 365: 545–48.

Winterbottom, M., T. Burke, and T. R. Birkhead, 1999. A stimulatory phalloid organ in a weaver bird. *Nature* 399: 28.

———, 2001. The phalloid organ, orgasm, and sperm competition in a polygynandrous bird: The red-billed buffalo weaver (*Bubalornis niger*). *Behavioral Ecology and Sociobiology* 50: 474–82.

Wolfner, M. F., 1997. Tokens of love: Function and regulation of *Drosophila* male accessory gland products. *Insect Biochemistry and Molecular Biology* 27: 179–92.

Wootton, R. J., 1971. A note on the nest-raiding behavior of male sticklebacks. *Canadian Journal of Zoology* 49: 960–62.

Wourms, J. P., 1977. Reproduction and development in chondrichthyan fishes. *American Zoologist* 17: 379–410.

Woyke, J., 1963. What happens to diploid drone larvae in a honeybee colony. *Journal of Apicultural Research* 2: 73–75.

Wrege, P. H., and S. T. Emlen, 1987. Biochemical determination of parental uncertainty in white-fronted bee-eaters. *Behavioral Ecology and Sociobiology* 20: 153–60.

Wrensch, D. L., and M. A. Ebbert, eds., 1993. *Evolution and Diversity of Sex Ratio in Insects and Mites.* Chapman and Hall.

Wynne-Edwards, K. E., 1995. Biparental care in Djungarian but not Siberian dwarf hamsters (*Phodopus*). *Animal Behaviour* 50: 1571–85.

Yamamoto, D., J.-M. Jallon, and A. Komatsu, 1997. Genetic dissection of sexual behavior in *Drosophila melanogaster. Annual Review of Entomology* 42: 551–85.

Yang, Z., W. J. Swanson, and V. D. Vacquier, 2000. Maximum-likelihood analysis of molecular adaptation in abalone sperm lysin reveals variable selective pressures among lineages and sites. *Molecular Biology and Evolution* 17: 1446–55.

Yeargan, K. V., 1994. Biology of bolas spiders. *Annual Review of Entomology* 39: 81–99.

Yeargan, K. V., and L. W. Quate, 1996. Juvenile bolas spiders attract psychodid flies. *Oecologia* 106: 266–71.

———, 1997. Adult male bolas spiders retain juvenile hunting tactics. *Oecologia* 112: 572–76.

Young, J. Z., 1959. Observations on *Argonauta* and especially its method of feeding. *Proceedings of the Zoological Society of London* 133: 471–79.

Young, L. J., R. Nilsen, K. G. Waymire, G. R. MacGregor, and T. R. Insel, 1999. Increased affiliative response to vasopressin in mice expressing the $V_{1a}$ receptor from a monogamous vole. *Nature* 400: 766–68.

Young, L. J., Z. Wang, and T. R. Insel, 1998. Neuroendocrine bases of monogamy. *Trends in Neuroscience* 21: 71–75.

Yund, P. O., 2000. How severe is sperm limitation in natural populations of marine free-spawners? *Trends in Ecology and Evolution* 15: 10–13.

Yusa, Y., 1996. Utilization and degree of depletion of exogenous sperm in three hermaphroditic sea hares of the genus *Aplysia* (Gastropoda: Opisthobranchia). *Journal of Molluscan Studies* 62: 113–20.

Zamudio, K. R., and B. Sinervo, 2000. Polygyny, mate-guarding, and posthumous fertilization as alternative male mating strategies. *Proceedings of the National Academy of Sciences, USA* 97: 14427–32.

Zann, R., 1994. Effects of band color on survivorship, body condition, and reproductive effort of free-living Australian zebra finches. *Auk* 111: 131–42.

Zeh, D. W., and J. A. Zeh, 1992. Dispersal-generated sexual selection in a beetle-riding pseudoscorpion. *Behavioral Ecology and Sociobiology* 30: 135–42.

Zeh, D. W., J. A. Zeh, and E. Bermingham, 1997. Polyandrous, sperm-storing females: Carriers of male genotypes through episodes of adverse selection. *Proceedings of the Royal Society of London, B* 264: 119–25.

Zeh, J. A., 1997. Polyandry and enhanced reproductive success in the harlequin-beetle-riding pseudoscorpion. *Behavioral Ecology and Sociobiology* 40: 111–18.

Zeh, J. A., and D. W. Zeh, 1996. The evolution of polyandry I: Intragenomic conflict and genetic incompatibility. *Proceedings of the Royal Society of London, B* 263: 1711–17.

———, 1997. The evolution of polyandry II: Post-copulatory defences against genetic incompatibility. *Proceedings of the Royal Society of London, B* 264: 69–75.

Zeh, J. A., S. D. Newcomer, and D. W. Zeh, 1998. Polyandrous females discriminate against previous mates. *Proceedings of the National Academy of Sciences, USA* 95: 13732–36.

# ACKNOWLEDGMENTS

Dr. Tatiana was born in Oliver Morton's kitchen, with Oliver himself, Jonathan Rauch, Anthony Gottlieb, Brian Barry, and Peter David attending as midwives. Without their help, she would never have come into being. She made her debut in *The Economist*. Many thanks to Bill Emmott for permission to draw on that first column for this book.

An enormous number of people have contributed to this book, whether by answering queries, helping hunt down obscure scientific papers, catching errors, commenting on parts of the manuscript, or simply letting me talk their ears off as I wrestled with ideas and arguments. For help with particular subjects or organisms, I would like to thank Phil Agnew, Matz Berggren, Philippe Bouchet, Stuart Butchart, Bill Cade, Tracey Chapman, Adam Chippindale, Andrew Cockburn, Bryan Danforth, Anne-Katrin Eggert, David Funk, David Gems, Darryl Gwynne, Peter Henderson, René Hessling, Rolf Hoekstra, Laurence Hurst, David Mark Welch, Nico Michiels, Christine Nalepa, Steve Palumbi, Charlie Paxton, Scott Pitnick, Heather Proctor, Bill Rice, Scott Sakaluk, Janet Shellman, Steve Shuster, Leigh Simmons, Mike Siva-Jothy, Donald Steinkraus, Willie Swanson, David Tarpy, Scott Taylor, Ethan Temeles, Barbara Thorne, Fritz Vollrath, Dave Walter, Stuart West, Martin Wikelski, Kenneth Yeargan, and Larry Young. The staff of several libraries were extremely helpful; but special thanks to the staff in the entomology library of London's Natural History Museum, who were tireless in helping me dig out information on obscure insects. Thanks, too, to Imperial College for the free run of their facilities. I am grateful to Ursula Mittwoch, who, in a letter responding to my column in *The Economist*, drew my attention to problems with Bateman's principle and prompted me to inspect the subject more closely. Thomas Bataillon, Austin Burt, Isabelle Chuine, Élodie Gazave, Philippe Jarne, Nicholas Judson,

ACKNOWLEDGMENTS 299

Thomas Lenormand, Armand Leroi, Yannis Michalakis, Ben Normark, Michel Raymond, François Rousset, and Denis Roze were regular and stimulating sounding boards; their suggestions and good humor were indispensable.

Peter Barnes, Thomas Bataillon, Bruce Greig, Greg Hurst, Ben Normark, Andrew Pomiankowski, Michel Raymond, Mark Suzman, and Stuart West kindly agreed to be guinea pigs for an early draft of part one; Austin Burt, Barbara Mable, and David Mark Welch read a draft of the final chapter; Caroline Daniel, Thomas Lenormand, François Rousset, Anthony Shewell, and Mark Suzman read the entire manuscript. Many thanks to all of them for helpful comments and vigorous criticisms. Bill Hamilton was the first to alert me to the problem of the evolution of sex, and to the extraordinary variety of sexual practices out there in nature. This book grew out of the work that I did with him during my doctorate, and I am sorry that he did not live to meet Dr. Tatiana.

Many thanks to Georges Borchardt and DeAnna Heindel for helping make this project a reality. To Alison Samuel and Penny Hoare at Chatto, and to John Sterling and Sara Bershtel at Metropolitan, for whom waiting for Dr. Tatiana must have sometimes felt like waiting for Godot, I owe particular thanks for encouragement and patience throughout. Many thanks, too, to Penny Hoare, Roslyn Schloss, Shara Kay, and above all to Sara Bershtel and her colleague Riva Hocherman, for comments and suggestions that greatly improved the manuscript.

Dr. Tatiana was blessed with an array of agony uncles: I am forever indebted to Dan Haydon, Horace Judson, Gideon Lichfield, and Jonathan Swire, all of whom read and reread draft after draft of chapter after chapter, gave suggestions that showed me paths through difficult thickets, and helped me break through bouts of writer's block, all the while providing boundless encouragement at all hours of the day and night.

I've been lucky to have had inspirational surroundings. I made most of the conceptual breakthroughs while staying at the Hôtel de l'Orange, in Sommières, France. Philippe de Frémont and his family welcomed a stranger, and provided a tranquil and beautiful sanctuary (and incidentally, improved my French). Jonathan Swire generously lent me Mosewell, an idyllic setting for writing. Sir John and Lady Swire, and Barnaby and Camilla Swire pampered the hermit who'd landed in their midst and gave me enormous encouragement and enthusiasm (not to mention, offering fabulous distractions that tempted me away from Dr. Tatiana for a few hours here and there). To Philippe and to the Swires, I cannot adequately express my gratitude.

Finally, I must thank Horace, for always being encouraging; Nicholas, for always laughing, and for plundering distant libraries for obscure papers on my behalf; my late mother, who taught me to play practical jokes (though she would've pretended to find Dr. Tatiana shocking); Mark, who rashly suggested I write a book about sex in the first place and had the misfortune to bear the brunt of the consequences; and Jonathan, for everything.

# INDEX

For specific species see the notes section.

# ABOUT THE AUTHOR

An alumna of Stanford with a doctorate from Oxford, OLIVIA JUDSON is an evolutionary biologist and award-winning journalist who has published in *The Economist, Nature, Science,* and *The Times Higher Education Supplement.* She is presently a research fellow at Imperial College in London. This is her first book. Visit Dr. Tatiana at her Web site: www.drtatiana.com.